How to Market Professional Design Services

How to Market Professional Design Services

GERRE L. JONES

McGRAW-HILL BOOK COMPANY

New York   St. Louis   San Francisco   London
Düsseldorf   Johannesburg   Kuala Lumpur
Rio de Janeiro   Singapore   Toronto
Montreal   Mexico   Sydney
Panama   New Delhi

Library of Congress Cataloging in Publication Data

Jones, Gerre L, date.
  How to market professional design services.

  1. Architectural services marketing.   2. Engineering
services marketing.   I. Title.
NA1996.J66   659.2'9'72   73-12032
ISBN 0-07-032800-5

234567890KPKP7654

*The editors for this book were Jeremy Robinson and Carolyn Nagy,
the designer was Naomi Auerbach, and its production was
supervised by George Oechsner.   It was set in Alphatype Astro
by University Graphics, Inc.*

It was printed and bound by The Kingsport Press.

Contents

Preface

The belief that an architect or engineer somehow debases his profession and himself by an active program of business development is no more relevant to today's mode of practice than covering the studio walls with carefully executed pencil sketches and water colors of Notre Dame, Sacré-Coeur, the Parthenon, and the Piazza San Marco. In spite of a professed revulsion to ethical promotion on the part of many professionals over the years, advice about the importance of maintaining a steady flow of new work into the office has been fairly plentiful. There is a popular (and probably apocryphal) remark attributed to at least a half-dozen former and present practitioners: "Young man, there are three things you need to know; get the job, get the job, and get the job." Such exhortatory but unhelpful counsel does little to advance the sometimes mysterious world of selling design services to potential clients.

Things really have not improved much in the more than thirty years since the late Royal Barry Wills wrote: ". . . nobody expounds the fundamentals of job getting or efficient office maintenance. By great good luck the archaic systems prevalent are usually kept under the architect's hat, which prevents their dissemination."

Up to now, then, the design professional traditionally has had to learn selling techniques essentially by experience, through trial and error, and with no real standards against which to measure the degree of his successes and failures. "Experience keeps a dear school," according to one of the maxims in Benjamin Franklin's *Poor Richard's Almanac,* but if one can find a few shortcuts and establish some guidelines along the way, his schooling becomes progressively less dear.

Just as the knowledge and skills required for the conduct of a responsible professional practice today are different from those of only twenty years ago, the necessary abilities and accepted techniques for marketing design services have undergone significant changes in the past few years. Change is no stranger to any of us, and since the chances are remote that it will ever occur in any form other than increasingly large doses, the important thing is how the design professional will react, adapt, and accommodate to future changes.

This book is a distillation of twenty-five years' experience in promotion and marketing in the United States and Europe. Some ten years of that quarter century have been spent in counseling design professionals—ranging from seven- to four hundred-man offices—on the subjects covered in the following pages.

No one, the author is convinced, has all of the answers nor productively utilizes a significant percentage of the known, proven ways to develop new business. However, a long-time exposure to the client acquisition activities of many firms—small, medium, and large—should offer a unique educational experience to the observer with an interest in chronicling their successes and failures—and the reasons for both. An impartial feedback of the information so gained forms the basis for many sections of this book.

Other parts of the book are based on material developed in recent years for seminars and workshops on successful marketing of professional design services. These sessions normally bring together participants representing practically all of the disciplines encompassed in today's total service team approach to development, design, and construction. Since these design professionals come from offices with staffs of five to over five hundred persons, a fairly wide and highly interesting cross section of problems, techniques, experience, and personalities emerges in retrospect.

This book, therefore, is partly history, partly philosophy, partly an exposition on how-to-do-it (based to some extent on how-others-did-it), and partly a look into the near future of the design profession. Since today's accomplishments are tomorrow's history, many readers of this book may be witting or unwitting contributors to its future editions, depending upon how well or how badly they put the principles outlined herein into effect in their own practices.

Gerre L. Jones

Introduction

Experience tells us that most readers of this book will fit generally into one of these four broad categories: architectural, engineering, and planning students; the principals of newly established design firms; educators in the design professions; and experienced executives of large, established firms. Individual readers may find themselves reflected in one or more of the following reader profiles.

I.

You are in your last year of engineering school. Since you had the good luck to have a classmate whose father is a partner in a medium-sized consulting firm in St. Louis, you have three summers of practical experience to your credit.

Last summer your boss did some griping about a large airport job that went on the shelf without warning. You also observed that other members of the firm occasionally got up tight over losing a job to another firm.

Because of these and other indications from your summer job that a steady flow of clients is important to the professional, you've

begun to wonder whether or not a couple of additional courses—
call them Business Development I and II—should not be part of
your school's curriculum.   Perhaps a few other practical courses
might also be considered—Introduction to News Writing, for
example, to assist the budding consultant in preparing understandable news stories about his jobs.   And Basic Public Relations,
for all of the unexpected problems which can arise in dealing with
clients, professional committees, the press, zoning and planning
commissions, and the public in general.   Survey courses in accounting, psychology, personnel practices, and business administration
should be equally helpful in preparing the professional-in-training
for a business career—what E. M. Forster called "the world of
anger and telegrams" and what the less imaginative call "the real
world."

Unfortunately, these extra classes could easily result in the addition of at least another year of study to the six or seven years of
college already required of architectural and engineering students.
Since few would seriously espouse that solution at this time, it is
our earnest hope that this book will serve as much more than a
primer for most of these untaught but patently desirable subjects.

Beyond this book, seminars and lectures on marketing professional services are offered periodically by certain management-
oriented organizations and associations.   Some of these courses
are excellent, many are mediocre, and a few are useless or worse.
The design professional's occasional attendance at the better organized sessions should be considered a valuable and practical
addition to his continuing education program.   They are tax deductible, of course, as is the cost of this book.

II.

You earned your professional degree a few years ago, served
your internship in a medium- to large-sized office, passed the
registration examination on the second try, and recently opened
your own office with two fellow practitioners who shared your
concern about stifling your creative design talent by working on one
multimillion dollar project after another in a large multidiscipline
office.

Practical fellows that you are, you signed a contract for a
$200,000 school addition before going on your own and all three of
you have a house design job or two in your pockets.

Work on the school addition is rolling right along; construction documents will be completed on schedule—and over coffee a few mornings ago, one of you hesitatingly suggested that you'd all better "look for some work." It was not until that moment that you recalled one of the advantages of working on those multimillion dollar projects in that too-large, insensitive office—someone else had always been responsible for bringing in the work.

This is probably as good a time as any to explain that several terms will be used interchangeably throughout this book. Business development, client acquisition, professional services marketing, client development—they all mean the same thing: selling your firm and its abilities to potential clients. Euphemisms abound in the design and consulting professions—in all professions, for that matter—and the man (or men) in charge of sales in your office may go by the title of Director of Client Relations, Executive Assistant for Business Development, Director of Communications, Consultant, or even by no title at all. We will have more to say on this point later; let's get back to our reader categorizations.

III.

As Chairman of the Department of Community Planning in the College of Design and Architecture in a Midwestern university you have seen your share of bright, ambitious young men and women graduate into the profession. Some of them return to your office from time to time, with news of themselves and other former students—and of their successes and disappointments. A few of the older men come back as visiting critics or to be interviewed by an architectural selection committee for one of the many new buildings going up on what has become a rather overbuilt campus. Of this latter group you know of a handful who have built up highly successful practices in the fifteen or twenty years since they left school. Some of your other graduates work for these men as draftsmen or spec writers or planners.

You have made it a part of your own continuing education to draw out the "winners" among your former students during these brief reunions. Some years ago you reached the conclusion that it requires a peculiar and uneven combination of designer-businessman-perfectionist-professional-administrator-promoter-egoist-Renaissance Man to become a standout in the design profession. Certainly not all of the top men have aspired to become "big," in the sense that their primary goal is to create a firm of hundreds of specialists, technical people, and special services groups.

What you would really like to be able to do, after all these years, is to impart to your students, before they graduate into the "dear school of experience," something of what makes the successful designers tick. Perhaps this book will be a partial answer for you and some of your students.

IV.

Unlike the readers already described, you have been in the profession long enough to have won some very large and significant commissions. Examples of your firm's work are to be found in thirty-eight of the fifty states and in a half-dozen foreign countries. You occupy a top slot in your firm's organization chart; you may well be a partner by this time.

You have seen offices come and go in the almost quarter-century of your professional career. Some, founded at the right time in the right place, took off like the proverbial skyrocket. A few years later they also fizzled out like the skyrocket—through bad or lackadaisical management, a refusal to grow with the times, the death of a key principal, hardening of the corporate arteries, or a combination of these and related front-office ailments.

Other offices, some owned by bright young men who started out with your firm, progressed steadily over the years to a position of professional eminence and financial stability.

You have not missed much in your own years on the board nor now behind an executive desk. You doubt that there will be much in this volume you have not already done or seen or heard about. Since your own experience has shown the inescapable correlation between learning and growth, and on the possibility that you may find a few helpful ideas—or at least affirmation and justification for your own philosophy about business development—you are willing to take a look at this book.

It is architects, planners, and engineers such as you to whom most of this book's less experienced readers will look for advice and help over the rough spots of their professional lives. We trust that you will indeed discover something of value in the following pages. Your comments and suggestions will be welcomed.

CHAPTER 1

Historical Background

Traditionally, such business-related terms and activities as "selling," "advertising," "marketing," "promotion"—even "public relations"—have been anathema to lawyers, physicians, architects, engineers, accountants, and certain other members of the learned professions. One has always been expected somehow to develop professionally and prosper fiscally with no external assistance and little, if any, personal promotion.

For the most part, architects and their fellow members of the design professions have been passive victims of this system, in that public patronage of the other professions is usually not an elective matter nor even up to individual choice of timing. When you are sick you see a doctor or perhaps die; if you have the misfortune to be arrested or sued an attorney is consulted because you do not want to go to jail or pay out large sums from your personal assets. Today the Internal Revenue Service makes an accountant a necessity, not a luxury.

On the other hand, the use of an architect, engineer, planner, or any of their numerous and related consultants normally is a matter of wide individual discretion. One can defer indefinitely the design and construction of an office building or a highway or an expansion of a high school. For proof of this look around at the vast numbers of obsolete hospitals, dangerous highways and bridges, overcrowded airports, run-down schools, and housing slums. And once the decision is finally made to build, as we are painfully aware, the potential client has an increasing number of alternatives to the retention of registered design professionals for his project.

PROFESSIONAL SHORTAGES

A separate but related problem for the architectural profession was pointed out several years ago by Bernard J. Grad, FAIA, in *Adventure into Architecture*. Even if all of the other problems were somehow magically resolved and clients were climbing over one another to shower significant commissions on every office, there still is the real danger of running out of manpower—competent and otherwise. According to Grad:

> There are only 30,000 registered architects in the United States, compared with 250,000 lawyers, 265,000 doctors, 430,000 accountants and 975,000 engineers. The number of architectural firms is about 10,000.
> The population of the United States is expected to reach 300,000,000 by the year 2000. To meet the physical demands of this expansion, during the next 32 years, architects will be called upon to design as many facilities as Americans have constructed in the 192 years since the signing of the Declaration of Independence.
> Only one in every 100,000 people is being prepared to participate in the physical shaping of the nation's future. The 63 accredited colleges of architecture conferred 2,025 Bachelor of Architecture degrees, 417 Master of Architecture degrees and only 10 Doctorates—less than 2,500 degrees in 1966. A pitifully small annual task force to meet the staggering challenge of the immediate future.[1]

While the numbers of graduates have increased since 1966, the rise has not been very encouraging to either educators or practitioners. Grad neglected to add that of the annual crop of 2,500 or so architectural graduates, a relatively large number enter other fields, including government, sales, and education. Probably no more than 60 percent of the newly graduated architects end up in one of the 10,000 architectural offices. Nor are all of the 30,000 architects currently registered actively pursuing their profession, by any means.

These figures mean that on the average there is one new graduate for every seven established firms to divide among themselves. The ratios for some of the other professions look like this:

One architect to every 8 lawyers
One architect to every 8½ doctors
One architect to every 14 accountants

Grad summarized his doleful statistics with the oft-heard complaint: "We are truly the unknown profession, not only because our number is small. To many we are simply the creators of blueprints. Others have absolutely no idea of what we really do in order to earn our fees. To nine out of ten clients and prospects, we are indistinguishable from the builders. This mystery must be dispelled if Americans wish to uproot the ugliness and visual disorder that now surround us."[2]

PROFESSIONAL CONFUSION

Dr. W. Graham Knox, M.D., then president of the Medical Society of the County of New York and writing in the April 1971 issue of *New York Medicine,* demonstrated that even other professionals can be pretty hazy about the architect's role. In cautioning his fellow Society members against overexposure to the news media, Dr. Knox commented:

> [More knowledge of the workings of our profession] should be allowed reasonably and intelligently, with no emphasis that one physician or one hospital possesses singular ability or obtains outstanding results exclusive of all others. The minutiae of the architect's drawings, the caisson worker's tricks of excavating, the submariner's or jet pilot's intricate maneuverings are kept pretty much from the public's inquisitive eye; why pronounce in detail the risks, complications and good and bad results attendant on modern advanced medical techniques?

Forgetting the doctor's gratuitous insult of coupling architecture with ditch digging, he seems blissfully unaware of the wide and constant exposure of "the minutiae of the architect's drawings to the public's inquisitive eye." Perhaps it would be instructive for a surgeon to be once required to file a full operation plan, including specs on the surgical tools and techniques he plans to use; the number of cc's of anesthetic to be administered; number, type, and size of sponges, etc., for an upcoming appendectomy on one of his patients with the local zoning board, plan commission, fine arts committee, and the patient's family. Hearings would be scheduled in advance of the operation, where the patient's neighbors and relatives could question the surgeon and his operating room team about their plan.   Naturally the newspaper, radio, and television reporters would want copies of the operation plan to publish far in advance of the surgery so that neighborhood groups in opposition to the plan could organize and less active members of the community could write letters to the editor.

As if small numbers and public confusion about his role were not enough to concern the architect, he seems increasingly to be running into the claims of other professions and disciplines.

The late Samuel B. Zizman, AIA, AIP, dealt at length with such misunderstandings and overlapping claims of professional competence in a paper for the 47th Annual Convention of the National Council of Architectural Registration Boards, meeting in Honolulu in July 1968.

> [The architect has] laid claim to any task, any project, any commission ennobled by the term of design. He has felt and claimed competence to design anything from a chair to a city. . . . Seemingly, he has claimed to be not only the master builder but master man.
>
> For a long time the engineer claimed that the art of city planning was his; now he turns a wary narrowed eye on the architect turned highwayman. The landscape architect has claimed that at least beginning with the building line, all the rest beyond is his domain. More recently, a host of others have staked out claims in the areas which once seemed sacred to the architect: the industrial designer is not content with packaging—he will design railroad stations, exhibition halls, and even city streets. The economist and sociologist at times push aside the problems of the spatial environment with the claims that until all of the figures are in, all the social problems are solved, all physical else is vanity. The architect has been challenged both in his treasured role as building designer and in his claims to work in the total physical environment.

We hardly need the reminders of these and many other spokesmen that all of the disciplines of the design profession, as a profession, require constant and continuous selling and reselling, and its members need all of the assistance they can get in promoting their individual practices. This book is designed to give the practitioner as well as the aspirant the benefit of successful, proven approaches to ethical client development. The book, it should be made very clear, is not an attack on professionalism. Neither is it an attempt to downgrade existing standards of ethical practice. Rather, the intent is to explain and define many of the tools and techniques of business development available to any office or individual, within currently accepted ethical standards. While it is believed that this is the first such attempt to make this type of specialized and detailed information and guidance available in book form, it is hoped that it will not be the last such effort.

EARLIER STUDIES OF BUSINESS DEVELOPMENT

A few authors have previously dealt with the subject of business development for the design professional, usually as one brief chapter in a book on the general subject of office practice or in such elementary terms that the result is little more than a primer for novices.

The first serious attempt to wrestle publicly with the question of how far the architect and engineer ought to go in seeking out clients was in a book published over three decades ago, *This Business of Architecture,* by Royal Barry Wills. Much of what Wills had to say about "this business of architecture" holds up surprisingly well more than thirty years later. His dedication, for example, was:

> To young men in architecture upon whose shoulders must fall the task of restoring the profession to its proper position of leadership, as much through the exercise of sound common sense in good business practice as though order and beauty in the design of their buildings.[3]

In his foreword Wills commented that "there have been architects for 5,000 years, yet the heritage of their accumulated experience does not spell the answer to success in the world of today. Why? Because the practice of architecture has been forced to take a course in business and to accept efficient, aggressive organization as the price of survival."[4]

COMPETITION VERSUS PROFESSIONALISM

Many architects and engineers are still exceedingly wary of anyone or anything which might appear to make their life's work out as a business rather than a profession. The idea that anything approaching "competition" exists within the profession is abhorrent on principle to this group of practitioners.

Your Chapter Guide to Effective Public Relations, a publication of the American Institute of Architects, intones: "Professional status is, in itself, a competitive advantage; it should not be thoughtlessly eroded." The identical statement, punctuated somewhat differently, appears in the chapter on public relations in the AIA's *Handbook of Professional Practice.*[5]

On this point Wills said: "The architect is still a professional and always must be; it is one of the strong arguments in his favor, but now his ancient lineage has also to be infused with the technic of a businessman. It is not an easy transition to make, and yet there is no other way to success amidst intense competition from within and without the profession."[6]

PRINCIPLES OF PROFESSIONALISM

Guidelines on the principles of professionalism were noted recently by Ralph E. Frede, Director of Development and Public Relations for the Baylor College of Medicine. Frede's seven statutes of professionalism are:

1. A profession must be dedicated to public rather than to personal interest.
2. A profession must have a distinct body of knowledge—information and skills used to serve the public good.
3. A profession must constantly expand its body of knowledge through systematic research.
4. The body of knowledge must be taught to apprentices who aspire to professionalism.
5. There must be examination and trial to establish competency—to protect employers, the public, and those engaged in ethical practice.
6. There must be an enforceable code of standards regulating performance.
7. It is incumbent upon professionals to continue their studies in order that they may better serve.

Paul B. Farrell, Jr., architect, attorney, and author, points out that the "comprehensive services" philosophy urged by many on the design professions—and either adopted or under serious consideration by an increasing number of firms—brings up an interesting question of potential interprofessional conflict:

Should an architect avoid creating potential conflicts of interest by pretending expertise in areas where, at most, he has cursory knowledge of the subject, and, is he in a conflict situation even if he is a multidisciplinary expert in the building process?

Perhaps all this discussion is irrelevant, since in the end it will not be the fine lines one or the other professional will draw around his territory that will lead to the success of a professional. Rather, it is the client, or, more exactly, the marketplace, comprising a stream of clients, who must be satisfied that the professional did in fact help solve his problem. To the client, academic discussions about the meaning of professionalism are relatively meaningless; he wants results.

The lack of practical courses in client acquisition in the universities was mentioned earlier. Royal Barry Wills had this to say in 1941: "Few architectural schools touch upon the subject and treatises on professional practice avoid it as the plague, so the burden of the teaching has been left to bitter experience, a dear teacher in the worst sense."[7] He added, "By great good luck the archaic systems prevalent are usually kept under the architect's hat, which prevents their dissemination."

One begins to feel that Mr. Wills, if years ahead of his time, was on the right track and it is to be regretted that he did not expand his ideas about business development into a complete volume instead of devoting a short chapter, "Stalking and Capture of Clients," to the fundamentals of job acquisition.

WILLS ON SELLING

In the chapter mentioned Wills gave a list of twenty-three suggestions for developing new business. Wills's practice was primarily limited to house design (another of his often whimsical books was about tree houses) and this practice limitation is reflected in his suggestions for obtaining clients. But because many of his points are still valid and will be discussed further and at length in following chapters of this book, we cite the whole list, which was headed "Indirect Selling by Individual Practitioners (to make clients come to you)":

1. Temporary exhibits in architects' exhibit space, stores, newspaper office windows or building report headquarters.
2. Newspaper and magazine articles on the practical and economic phases of building, or popularized stories about architecture, or on momentarily controversial subjects relating to architecture.
3. Regular appearance of sketch plans and elevations in newspapers.
4. A sequence of printed outlines, covering the nature and value of architectural services and the advantages of supervision, to a selected mailing list.
5. A folder of sketch plans and elevations, displayed in such places as doctors', oculists', or dentists' waiting rooms.

6. Illustrated lectures of talks before clubs.

7. Movies, "dramatizing" your previous work.

8. Architectural competitions.

9. Exhibition houses.

10. Civic service in town, city planning, or art commissions.

11. Real estate brokers' recommendations.

12. Recommendations from friends or club or fraternal acquaintances.

13. Spreading goodwill by "disinterested" bits of architectural advice, whenever the opportunity arises.

14. Having a wife in club and civic work, who keeps her ear to the ground.

15. Making the most of your college alumni connections.

16. Printed appeal through a combination service—financial, land, architectural.

17. Recommendation of satisfied clients.

18. Regulated social activity, not wasted in the charming company of your own competitors.   Hard-boiled but effective.

19. Getting your name in print, as John Brown, architect, for any worthy reason.

20. Going to church more often than Easter and Christmas.

21. Never avoiding a friendly conversation with an apparently solvent person, even though he be a stranger.

22. Giving a "university extension course" relating to architecture.

23. Sending a photostated montage of documents attesting to your successes and satisfied clients to where it will do the most good.[8]

Several of his recommendations, because they appeared to skirt or even ignore AIA ethical standards, no doubt caused a few raised eyebrows among Wills's pre-World War II fellow professionals.

1909 CANONS OF ETHICS

It may come as somewhat of a shock to many of the AIA's current members to learn that organized architecture somehow muddled along for more than fifty years without a formal set of rules governing professional practice and ethics.   Not until the Washington, D.C., convention in December 1909 did the AIA adopt and promulgate the first Canons of Ethics, along with *A Circular of Advice Relative to Principles of Professional Practice*. This initial approach to defining standards of ethical practice was first published in the 1910 *AIA Annuary,* a kind of predecessor of *The Journal of the AIA.*

The story goes that the AIA was finally moved to spell out standards of professional practice and conduct because of a proliferation of unauthorized architectural competitions. In this connection Principle of Professional Practice number 14, "On Competitions," said:

> An architect should not take part in a competition as competitor or professional advisor or juror unless the competition is to be conducted according to the best practice and usage of the profession as formulated by the Institute.   Except as an authorized competitor, he may not attempt to secure work for which a competition has been instituted. He may not attempt to influence the award in a competition in which he has submitted drawings.   He may not accept the commission to do the work for which a competition has been instituted if he has acted in an advisory capacity either in drawing the programme or in making the award.

There were eighteen other "Principles," covering such matters as fees, engaging in building trades, encouraging good workmanship, advertising, injuring or supplanting other architects, and professional qualifications.

SITE SIGNS

One of the original principles, number 13, apparently fell unmarked and unmourned along the wayside.   Titled "On Signing Buildings and Use of Titles," it cautioned the architect that "the display of the architect's name upon a building under construction is condemned, but the unobtrusive signature of buildings after completion has the approval of the Institute."

Compare the 1909 Principle with the section on "Site Signs" in "The Architect and Public Relations," Chapter 8 of the *Architect's Handbook of Professional Practice.*[9]

> SITE SIGNS.   Site signs offer a very effective and inexpensive method of communication which every Architect can use to inform the public of his involvement with a project.   This can be done through cooperative efforts in creating one well designed site sign that lists the title of the project and those involved in design, construction, financing, etc.   In some cases, it may be appropriate to include on the sign an architectural drawing of the new building, but the artwork should be of outstanding quality.   Generally, personalized, individual site signs can be created and displayed by the architect. . . .

Of perhaps incidental interest, in 1961 the Royal Institute of British Architects—from whose current *Journal* editor we will hear more at the end of this chapter—proposed a standard job sign for the use of its members. The sign, in two sizes, featured the RIBA seal in the left quarter, the architect's name and address in the center half, and the word "Architects" appearing in reverse (white on black) on the right quarter. The signs were available from designated sign makers. RIBA members were not required to use the official sign but were prohibited from using the RIBA seal on signs of their own design.

A comparison of the Canons of Ethics which accompanied the *Circular* with the current AIA Standards of Ethical Practice may be instructive or at least of some historical interest.

The Canons of Ethics (1909)

The following Canons are adopted by the American Institute of Architects as a general guide, yet the enumeration of particular duties should not be construed as a denial of the existence of others equally important although not specifically mentioned. It should also be noted that the several sections indicate offences of greatly varying degrees of gravity.

It is unprofessional for an architect—

1. To engage directly or indirectly in any of the building trades.

2. To guarantee an estimate or contract by bond or otherwise.

3. To accept any commission or substantial service from a contractor or from any interested party other than the owner.

4. To advertise.

5. To take part in any competition the terms of which are not in harmony with the principles approved by the Institute.

6. To attempt in any way, except as a duly authorized competitor, to secure work for which a competition is in progress.

7. To attempt to influence, either directly or indirectly, the award of a competition in which he is a competitor.

8. To accept the commission to do the work for which a competition has been instituted if he acted in an advisory capacity, either in drawing the programme or in making the award.

9. To injure falsely or maliciously, directly or indirectly, the professional reputation, prospects or business of a fellow architect.

10. To undertake a commission while the claims for compensation, or damages, or both, of an architect previously employed and whose employment has been terminated remains unsatisfied, until such claim has been referred to arbitration or issue has been joined at law, or unless the architect previously employed neglects to press his claim legally.

11. To attempt to supplant a fellow architect after definite steps have been taken toward his employment.

12. To compete knowingly with a fellow architect for employment on the basis of professional charges.

AIA STANDARDS OF ETHICAL PRACTICE

The 1909 Canons have been superseded in all areas by the *Standards of Ethical Practice* (AIA Document J330, Revised November 1, 1970). A careful comparison of the two documents, adopted some sixty years apart, shows that all of the 1909 Canons are essentially covered by the 1970 Standards with one exception; the Canons included a stricture against guaranteeing an estimate or contract by bond or otherwise. This omission in the later document may be accounted for by the fact that in today's market the client expects that his budget will be honored—and the designer who consistently treats estimates and budgets lightly will not be in business very long.

The principal of at least one large architectural firm makes it a practice during presentations to potential clients to guarantee that the building will be bid within 5 percent of the final estimate, or the job will be redocumented at the architect's expense. Most clients do not readily pick up the point that this guarantee gives the estimators and the architect a 5-percent deviation either way, or a total allowable variance of 10 percent.

This same architect usually follows up his guarantee with the remark that his record over the years averages out to less than a 2-percent difference from final estimates. One danger in making such statements is that the client is more apt to remember the 2-percent average figure achieved over twenty years than the 5-percent guarantee.

That four of the 1909 Canons dealt with architectural competitions (numbers 5, 6, 7, and 8) lends credence to the story that competition abuses prompted the AIA to formalize professional practice standards. Of the 1970 Standards, numbers 1, 2, 4, 5, 8, and 11 do not appear to have been covered in the older Canons. These deal with improving the human environment, support of human rights, endorsements of products and services, preserving confidences, contributions to secure commissions, and encouraging education and research.

NSPE CODE OF ETHICS

Still another interesting difference in the two versions is that the early Canons were couched in essentially prohibitive or negative terms, while the modern Standards are stated in a rather positive—even permissive—manner.

As might be expected, the various codes of ethics, standards of ethical practice, and mandatory standards of all organizations involved in the design professions do not vary significantly from one another as regards business development.   The Code of Ethics of the National Society of Professional Engineers, for example, covers approximately the same points as the AIA's Standards of Ethical Practice, but in considerably greater detail.   The ability of the engineer to advertise is one of the primary differences in the two codes.   Section 3a(1) of the NSPE Code[10] reads:

> The Engineer shall not advertise his professional services but may utilize the following means of identification:
> Professional cards and listings in recognized and dignified publications, provided they are consistent in size and are in a section of the publication regularly devoted to such professional cards and listings. The information displayed must be restricted to firm name, address, telephone number, appropriate symbol, name of principal participants and fields of practice in which the firm is qualified.

On professionalism the NSPE says:

> Professionalism and ethics are twins, inseparably bound together in the concept that professional status and recognition must be based upon public service under a higher duty than mere compliance with the letter of the law. . . .   A Code of Ethics is not a static document; its purpose is to live and breathe with the profession it serves.   Experience and changed circumstances will require continual review and revision in order that the Code of Ethics will also reflect the growing understanding of engineering professionalism in public service.[11]

Following a month-long visit to the United States in early 1971, Malcolm MacEwen, editor of *The Journal of the Royal Institute of British Architects,* made some interesting observations on the AIA's 1970 Standards of Ethical Practice in the February 1971 issue of the RIBA *Journal.*

> The split personality of the [architectural] profession [in the United States] is clearly reflected in the revised code of ethics.   The real effect of the code is not to promote public interest but to weaken the restraints on commercialism.   It is couched in such woolly generalities that, apart from a ban on advertising, it permits the private practitioner to do more or less as he pleases—provided he does it with a good conscience. Nobody believes that many architects will in fact refuse commissions because their rich corporation clients are damaging the environment.

It seems improbable that the AIA will be able indefinitely to perform the balancing act which enables its practicing members to proclaim their devotion to the concepts of human environment and human rights in one breath, and to accept whatever commissions come along in the next. They have increasingly to get out and get them, to enter into property speculation, to become subsidiaries of conglomerate corporations or to ally themselves with financial interests of one kind or another, if they are to survive.

Editor MacEwen's trenchant summation of the state of American architecture—indeed, of the entire design profession—might well serve as the primary theme of this work. Others of far greater competence in the fields of social obligations, human rights, and environmental concerns have written widely and wisely of the profession's shortcomings in those areas. This book, then, will concern itself with how to "get out and get them" as Mr. MacEwen described the search for new business; in a word, survival.

REFERENCES

[1] Bernard J. Grad, *Adventure into Architecture,* Arco, New York, 1968, p. 11.

[2] Ibid.

[3] Royal Barry Wills, with collaboration of Leon Keach, *This Business of Architecture,* Reinhold Publishing Corporation, New York, 1941.

[4] Ibid.

[5] *Architect's Handbook of Professional Practice.* Excerpted from chap. 8 with permission of the American Institute of Architects. Washington, D.C., 1971.

[6] Wills, op. cit.

[7] Ibid.

[8] Ibid., pp. 37–38.

[9] AIA, op. cit.

[10] *Code of Ethics for Engineers,* NSPE Publication No. 1102 as revised, Washington, D.C., January 1971.

[11] *Ethics for Engineers,* NSPE Publication No. 1105, Washington, D.C., May 1970.

CHAPTER 2

Principles and Psychology
of Marketing

Alert readers will realize that the title of this chapter is also the principal theme of this book, *How to Market Professional Design Services;* indeed, the book subject conceivably could be a subtitle of this chapter heading. We are not really going to be able to cover the broad subject of "Principles and Psychology of Marketing" in a few pages, and those who want to dig more deeply into the matter may assemble their own bibliography from the citations in the footnotes and from other readily available sources.

DEFINITIONS

With the disclaimers out of the way it might be helpful to begin with a few definitions:

Marketing, sometimes called "distribution," is the performance of business activities connected with the movement of goods and services from producers to consumers or other users. In addition to the analysis of these activities, marketing involves the comprehension of consumer circumstances and attitudes that determine the character of a major part of marketing activities; the business organizations that perform these activities; and relevant aspects of government regulation. In marketing, the ability to recognize early trends is fully as important as knowledge of the current state of affairs.[1]

Selling . . . is the personal, oral presentation of products or services to prospective customers for the purpose of making sales. [In the United States] it has become a highly developed technique, based on psychological analysis and psychological application. . . . The well-schooled salesman of today bases his presentation on his understanding of the customer's buying motivations as related to the particular product and by subtle stimulation and manipulation of these motivations the customer is led to want to buy the product. . . . The salesman ideally does more than make the customer desire the product; he tries to win the customer's regard for the company which sells the product [and] tries to extend the confidence and regard of the customer to himself.[2]

If we substitute "design professional" for "salesman" and "services" for "product" the definition of selling becomes directly related to the marketing of professional design services.

EFFECTIVE COMMUNICATIONS

Bound up in any definition of marketing and its mix of sales promotion, advertising, publicity, and public relations is the correlative requirement of effective communications.

Communication, simply stated, is the transmission of ideas from a speaker to an audience. The speaker sometimes is referred to as the "source," the transmission, the "signal," and the audience, the "recipient." The audience or recipient may be

An individual

A single audience of several hundred persons

A noncohesive or cellular audience numbering in the millions

Examples of these three audience types would be (1) a husband and wife conversation, (2) a talk on cost controls at an AIA regional conference, and (3) the total readership of a best-selling novel.

THE COMMUNICATIONS PROCESS

In their excellent book, *Promotional Strategy*, Professors Engel, Wales, and Warshaw explain the communication process as beginning

. . . with some event which stimulates Mr. A to transfer his ideas or notions to Mr. B. He selects certain words which he then arranges in a pattern or sequence to be communicated to a recipient. The process of selecting and arranging the words is called "encoding." The encoded words are then transferred to Mr. B through some kind of signal, perhaps spoken or written language. When Mr. B receives the signal he searches for meaning (decoding) by comparing the signal against his accustomed thought patterns. Communication takes place when Mr. B encodes his reply and transfers it through signals to Mr. A.

The message or signal is transmitted in the form of "symbols," which are nothing more than substitutes for the things they represent. A symbol acquires a more or less unique meaning through social consensus, and hopefully it will call forth the same response from both Mr. A and Mr. B when they are communicating.[3]

Suppose you are at a dinner party with a number of friends. The salad course has just been served, and a quick taste tells you that it needs salt. (This initial source-signal-recipient sequence involves only one individual—you—unless it can be argued that an inanimate object—the salad—may be a source.) You turn to the man on your left and say, "Please pass the salt." He answers, "Of course," and reaches for a saltshaker, picks it up, and passes it to you. This common situation contains all the elements of any communication. The first step was the formulation of your idea; you wanted the salt. Words were then selected (encoded) to convey your request and transmitted or transferred to your tablemate. He received and decoded your signal and responded by passing the saltcellar to you.

Formulation or encoding of ideas and their transmission normally are within the control of the speaker, but unless the ideas are formulated and transferred so as to gain the desired response, the procedure is ineffective and communication fails. If, in the example above, your dinner companion had passed you the pepper instead of the salt, his decoding was faulty and you would have had an obvious breakdown in communications.

QUESTIONS FOR A SALESMAN

Let us consider the following three questions:
Is anybody there?
Is anybody listening?
Does anybody care?

Those readers fortunate enough to have seen the Broadway musical *1776* will recognize those plaintive inquiries from one of General Washington's periodic messages to the Continental Congress sitting in Philadelphia. Washington was reduced to such desperation after many previous requests for arms and materials for his armies had been ignored or shrugged off by the Congress.

A design professional has not had much experience in selling his services to clients, either in preliminary one-to-one sessions or in full-fledged formal presentations, if he has not occasionally asked himself:

Is anybody there?

Is anybody listening?

Does anybody care?

SELL THE SIZZLE

Elmer Wheeler, a salesman's salesman of the 40s and 50s, summed up his understanding of marketing psychology in the catch phrase, "Don't sell the steak — sell the sizzle." Sizzle, according to Wheeler, appeals to the heart, while steak appeals to the brain. Sizzle sells faster and better because the pocketbook is closer to the heart than the brain.[4]

Another Wheelerism: "Always give the buyer a choice between something and something — never between something and nothing."[5] Related was his "Don't ask if — ask which," illustrated by the dentist asking, "Shall I fill the tooth now or wait until it hurts more?" and the doctor's inquiry, "Shall we go to the hospital now while you can still walk, or do you want to wait until the appendix bursts some night and you have to go in an ambulance?"[6]

A good salesman, by necessity, operates on many levels. The common-sense level dictates that he learn the prospect's viewpoint and adapt to it. The vice president of marketing for a large, successful U.S. corporation expressed his ideas on the importance of relating to a prospective client: "I believe that all of the other qualities of a true salesman are of lesser importance. He can have enthusiasm, stamina, intelligence, personality, sincerity and all of the other attributes of a salesman and still be a failure if he does not have the knack of finding the points of common interest. The important points are those which are of chief interest to the prospect — not to the salesman."

One could do worse, as unassigned homework for this chapter, than to dip into a few of Elmer Wheeler's books. In addition to the already cited *How to Sell When Selling is Tough,* you might scan

How to Make Your Sales Sizzle in 17 Days[7] and *Tested Ways to Close the Sale.*[8]

THE SELLING PROCESS

There are literally dozens of good books about marketing and selling, of which we will be able to highlight only a few.   In *The 5 Great Rules of Selling* Percy Whiting sets out the major steps of the selling process as attention, interest, conviction, desire, and close. "Somehow or other," Whiting says, "the purchaser of any item more important than a package of gum or a pack of cigarettes goes through these steps in making a purchase."[9]

Much of the literature about marketing is oriented to the selling of products rather than services.   However, much of what is written about the marketing of products applies equally well to selling professional services.

Steven Morse, in *The Practical Approach to Marketing Management,* breaks down the selling process (he calls it "sales confrontation") into seven steps, which are possibly more closely related to the selling of professional design services than some of the previous references:[10]

Preparation
Identification of needs
Presentation of benefits
Factual evidence
Overcoming objections
Taking the order
Analyzing the result

Morse's fifth step, overcoming objections, might also be stated as how to take the "but" out of "rebuttal."

SELLING BY DEMONSTRATION

A persuasive presentation of benefits may take many forms in selling.   Henry Ford, who is credited with the statement, "The way to a fortune is simple—find a public need and fill it," is also generally regarded as the man who developed the interchangeability of parts principle for his automobile assembly lines.   However, it was Eli Whitney of cotton gin fame who actually developed the concept—a century before the Model T made its debut.

Whitney was given his chance to demonstrate accurate machine production of interchangeable parts when the United States believed itself on the verge of war with France and in need of thousands of muskets. Rather than following the traditional process of gun making, wherein gunsmiths fabricated each weapon by hand and fitted the parts into a custom gun, Whitney was certain his machines would do the job much faster and more cheaply.

To make his point in a memorable and dramatic way Whitney took several completely disassembled guns to his meeting with the Army. Spreading the parts helter-skelter on a long table, he quickly assembled a musket from parts taken at random from the several piles, astonishing the military observers and winning the contract for 10,000 new muskets.

It is interesting to contemplate how much more commonly fame has resulted from what men said than from what they did. Theodore Levitt makes this point in *The Marketing Mode:*

> Thomas Jefferson's authorship of the Declaration of Independence did not keep him from maintaining a house full of slaves. Captain Lawrence got into folk history because only the first part of his story was told. Less than 15 minutes after he uttered his immortal words, "Don't give up the ship," he surrendered. Stonewall Jackson got his famous nickname from a later misinterpretation of what General Bernard Bee said at the first battle of Bull Run. Bee called for Jackson's assistance, but Jackson hesitated, and Bee complained: "Look at Jackson; he stands there like a stone wall."[11]

TEN PRINCIPLES OF SELLING

For those who prefer their selling principles reduced to lists, *Ten Ways to Get People to Respond to Your Sales Message* offers good advice in capsule form:

1. Find out what they want and need (remember Henry Ford's statement a few paragraphs back) and show them how you can help them. And do it honestly, too! When a sales message is delivered the listener has one predominant thought in mind—"What's in it for me?"

2. Prepare a "Want and Need Questionnaire" to help determine the interests of a sales prospect. There is no sense in trying to sell your organization, product, or services if you have not found out the wants and needs of the person you are trying to convince. He just won't listen.

3. Prepare a list of all the products and/or services of your organization. Concentrate on all the unique benefits you offer. Then show how your products, services, and unique benefits will fulfill the wants and needs of the person to whom you are delivering the sales message.

4. Correlate the needs and wants of the prospect with the products, services, and unique benefits you offer. This will help you arrive at a Unique Marketing Proposition for your organization. If you offer what everyone else offers the prospect will not listen. If you offer something unique in terms of his interest he will start to listen. (Ellerbe Architects, for example, describes its complete range of programming, planning, design, consulting, development, and construction services as "More than Architecture.")

5. Develop a unique Selling Idea or Theme and communicate it through your salesmen, in sales letters, general brochures, product brochures, newsletters, direct mail, annual report, packaging, advertising, billboards, point of purchase displays, exhibits, slide films or movies, or any other media.

6. Develop a unified graphic look around your selling theme, wherever it appears. It should be consistent and easily identifiable. Use all the principles of corporate identity to achieve this look—a look of quality that is distinctive and attention getting.

7. Be honest and straightforward in everything you do. People will stop listening if you are not. Especially in the 70s. The younger generation and the older one, too, are going to be far more difficult to convince unless you are straight with them.

8. You must really care about the prospect—and show it in every way. The more care and interest you honestly show, the more effectively your message will come through.

9. Your message must be repeated again and again because people are skeptical, even though you want to help them. The more you repeat your message, in different ways, the more it will be believed.

10. Demonstrate not only with graphics and words that you want to help a prospect but with concrete action. Show in as many concrete ways—case histories, etc.—as possible how a prospect will benefit if he responds to your message. If all things are equal between two companies—equal products, services, quality, and prices—the company that uses these ten principles will communicate much better than the one that does not.[12]

PERSUASIVE COMMUNICATION

As we have seen, one of the principles of effective marketing, around which a large measure of the psychology of marketing revolves, is persuasive communication. Even though we usually are aware of personal shortcomings in coping with what some psychologists call the I.T. Factor—Information Transmitted—we tend to forget the discouraging statistics of retention. About 20 percent of a spoken message is retained, on the average; some 30 percent of purely visual intake is retained, and a combination of aural and visual messages raises the retention factor to 50 percent.[13] We generally know the limitations of our own memory and powers of concentration—how frequently we "tune out" on a speaker—but we expect our audiences to have a limitless attention span and a motivation to listen to and absorb everything we say. The other person (audience) is also credited with the innate ability to put exactly the same interpretation on our words as we understand and mean them, with no allowance for misunderstanding or ambiguity.

On the other hand, we often and unrealistically consider that the other person speaks from a well-based, organized position, even though we are aware that our own ideas are usually formulated as we talk.

Dr. Jesse S. Nirenberg, a consulting industrial psychologist in New York City, outlined a six-part method of persuasive communication:

1. Marshalling of facts and working out of quantifications beforehand to support your position.
2. Motivating the other person to listen.
3. Matching the pace of your input to his receiving capability.
4. Getting feedback to stimulate his thinking about your input and to determine what additional input he needs.
5. Exploring the other person's objections with him to examine his premises, in order to help him develop a valid position.
6. Minimizing the other person's defensive and ego-inflating distortions of arguments.[14]

COMMUNICATION COMMANDMENTS

Dr. Nirenberg also points out that "communicating to gain the acceptance of an idea requires a high degree of the same characteristics that we evaluate by analyzing the other person's method of communicating: effort, initiative, objectivity, and identification with the needs of others—plus patience."

These four characteristics—effort, initiative, objectivity, and identification—are further explored as "communication commandments":

Effort: As indicated by his communicating, how much effort did the individual put into developing the information he is communicating? To what extent has he researched to get needed facts? How much analyzing of the facts has he done in order to come to his conclusions? How much effort does he put into trying to make you aware of his facts, his reasoning and his conclusions?

Initiative: To what extent does he ask questions to draw out information? Does he search for alternatives if the first way doesn't work? Does he work independently rather than depend on you to guide him? Does he volunteer information or do you tend to draw it out of him?

Identification (with the needs of others): To what extent does he explore your thinking in order to relate his ideas to your concerns? Does he focus on the benefits of the ideas he is presenting so as to motivate you to buy these ideas? Does he communicate the fact that he is absorbing what you are saying? Does he give you an opportunity to finish expressing your ideas rather than interrupt? Does he stick to the point or does he tend to wander into irrelevancies and to overelaborate as a way of complaining or reflecting some credit on himself?

Objectivity: To what extent does he bring in facts and figures to substantiate his case? Does he look at the opposing side as carefully as he looks at his own? Is he as open to criticism of his weak points as he is to praise of his strong ones? Does he accept or reject opinions on the basis of the individual expressing them and on wishful thinking, or does he base his judgment of the opinions on the facts behind them? Does he stretch the truth when it serves his purpose?[15]

COMMUNICATION FAILURES

The further a communication is removed from its primary source, the weaker the strength and authority of the communication becomes. Communications passed through secondary sources invariably become changed, diluted, and obscure—losing their clarity, purpose, validity, *and* persuasiveness.

An old party game involved the relaying of a simple message through a line of participants, on a one-to-one basis. After the message had progressed through a dozen or so players any resemblance between what the last person heard and what the originator started through the line was purely coincidental.

A favorite stunt in first-year journalism classes, to demonstrate the fallibility of prime sources, rather than secondary ones, has several strangers rush to the front of the classroom carrying on an argument among themselves in loud tones, punctuated by dire threats.   Suddenly, one of the protagonists pulls out a gun and shoots one of the others at point-blank range.   The perpetrator and his companions then rush from the room, followed by the shooting victim who has been miraculously restored to life.   At this point the professor asks the stunned students to write down the essential details of what they have just witnessed—the number of participants in the episode; their sex, race, dress; what type of gun was used; how many shots were fired—the barest of details.   Anyone who is familiar with the confusion of eye witnesses at a police showup will know how poorly the students do in trying to describe the scene they saw only moments before.   It is always a sad and discouraging commentary on human powers of observation—and a dramatic demonstration of failure in communications.

How can we guarantee against failures in communications? Unfortunately, there is no pat answer.   Semanticists and psychologists have long studied the question but have arrived at few conclusions and no real solutions.   We do know that effective and persuasive communication appears to involve empathy, with the speaker and the audience placing themselves in each other's shoes, in effect.   This is also called role taking, with at least three primary requirements for it to occur:

1. Some commonality of background and experience.

2. Some indication (feedback) from the audience that he is decoding the speaker's signal; a smile, an affirmative reply, a frown, or even no response at all helps the speaker determine whether or not the recipient is "getting the message."

3. The speaker must have some knowledge of his audience's motivational influences at the time of transmission.

CLARITY IN COMMUNICATIONS

Some speakers appear almost to search for ways to avoid communicating with their audience—a planned communications gap, as it were.   In proposals and presentations the design professional cannot afford any approach which does not produce clear writing and understandable speaking.   Do not emulate the Chinese government specification writer who asked for bids on manually operated biquinary computers when he meant abacuses; nor the U.S. Army specification man who requested prices on aerodynamic personnel decelerators (he wanted parachutes).

A few years ago the New York City Department of City Planning was worrying about "the alarming possibility that planners are seldom completely understood by the public." The department's newsletter for February–March 1967 carried a thoroughly tongue-in-cheek article, "Closing the Communications Gap." A few excerpts:

> [Planning] language is peppered with technical names for ordinary activity. Driving through a neighborhood and looking around is a "windshield survey in the field." All residents who move from the city are "gross out-migrants."

Even while attempting to close the communications gap, it is recognized that there are advantages to innocence. Consider the following excerpt from an internal report:

> Acceptance of the postulate framework and its resultant conceptualized statement diagramming the functioning of the education system within the community leads to an analysis of the components of the system as well as of the potential impacts and implications of the consequences of the process. This analysis is both prerequisite and past of the forma-tion of a new methodological approach which is an objective of this work. This is not to imply, however, that the entire system as shown in Diagram One is of equal weight in any given analytic situation. There are some aspects, whether because of their intrinsic value, or their extrinsic value as among all factors of the system, which must be dealt with in a cursory fashion.

The uninitiated may dismiss this as gobbledygook, the news-letter editor comments; however, the professional can only shift from one foot to the next, in awe, assuming that it must mean something, as indeed it does, probably.

It has always seemed the height (or depth) of irony that the English language word to describe a person who abhors the use of long words is "ultra-antihypersyllabicsesquipedalianist." It may destroy a few childhood illusions to point out that ultra-antihyper-syllabicsesquipedalianist beats out that old favorite of spelling bees, antidisestablishmentarianism, by eleven letters.

GOBBLEDYGOOK AND GOVERNMENTALESE

The Wordsmanship System Chart was invented for those persons who are fascinated with all forms of gobbledygook, governmentalese, and similar aberrations of the English language.   The original Wordsmanship method, also known as the Systematic Buzz Phrase Projector, was developed by a Washington, D.C., civil servant several years ago.   Employing a lexicon of thirty carefully chosen "buzzwords," it looked like this:

Column 1	Column 2	Column 3
0. integrated	0. management	0. options
1. total	1. organizational	1. flexibility
2. systematized	2. monitored	2. capability
3. parallel	3. reciprocal	3. mobility
4. functional	4. digital	4. programming
5. responsive	5. logistical	5. concept
6. optional	6. transitional	6. time-phase
7. synchronized	7. incremental	7. projection
8. compatible	8. third-generation	8. hardware
9. balanced	9. policy	9. contingency

The procedure is simple.   Think of any three-digit number, then select the corresponding buzzword from each column.   For instance, number 257 produces "systematized logistical projection," a phrase that can be dropped into virtually any report with that ring of decisive, knowledgeable authority so necessary to successful technical writing.

A few years later, architect R. Jackson Smith converted the buzzword chart to an "Architectural Innovator System," or "1,000 Guideposts to the Patter-Oriented-Process for Sloganized Architecture."   Herewith, Smith's P-O-P system:

Directions

1. Choose any three-digit number.
2. Find the word corresponding to each digit.
3. The three words should indicate an architectural innovation.

Column 1	Column 2	Column 3
1. architectural	1. engineered	1. designs
2. team	2. articulated	2. planning
3. computer	3. coordinated	3. projects
4. field	4. correlated	4. environment
5. behavior	5. modulated	5. schematics
6. group	6. directed	6. processes
7. resource	7. integrated	7. systems
8. realistic	8. controlled	8. concepts
9. value	9. oriented	9. structures
0. creative	0. programmed	0. assemblies

Example: 007 produces not James Bond but the slogan "creative programmed systems." This should suggest an architectural innovation; if it doesn't, try another number.

In summarizing this chapter we might draw on Victor Hugo's *The History of a Crime,* wherein he wrote, "More powerful than armies is an idea whose time is come." Perhaps slightly more relevant to the principles and practices we have discussed here is the remark attributed to an executive of the American Telephone and Telegraph Company: "Silence is still the best substitute for brains, though it is not yet an absolute replacement." Or we could fall back on the ninth of the nine deadly sins of marketing— "Promote what you want to design, rather than what the market wants to build." However, exercising an author's prerogative, we will close with the explanation of his business role by a partner in a New York brokerage firm: "I don't sell. People buy from me."

REFERENCES

[1] Reprinted with permission from volume 15 of *Collier's Encyclopedia,* p. 416. © 1971 Crowell-Collier Educational Corporation.

[2] Ibid., p. 422.

[3] James F. Engel, Hugh G. Wales, and Martin R. Warshaw, *Promotional Strategy,* Richard D. Irwin, Inc., Homewood, Ill., 1967, pp. 12–13.

[4] Elmer Wheeler, *How to Sell When Selling Is Tough,* Doubleday & Co., Inc., Garden City, N.Y., 1958, pp. 19–20.

[5] Ibid., p. 32.

[6] Ibid., pp. 32–33.

[7] Prentice-Hall, Inc., New York, 1953.

[8] Harper & Brothers, New York, 1957.

[9] Percy H. Whiting, *The 5 Great Rules of Selling,* rev. ed., McGraw-Hill Book Co., Inc., New York, 1957, p. 49.

[10] Steven Morse, *The Practical Approach to Marketing Management,* Mc-Graw-Hill Publishing Co., Ltd., London, 1967, p. 211.

[11] Theodore Levitt, *The Marketing Mode,* McGraw-Hill, Inc., New York, 1969, pp. 218–219.

[12] Courtesy of Corporate Image Planners, Philadelphia, Pa.

[13] George R. Snell, *Conspectus on Communications,* George R. Snell Associates, Mountainside, N.J., 1965, p. 1.

[14] Jesse S. Nirenberg, "The Hidden Tools in Persuasive Communication." Excerpted from the June 1971 issue of *Business Management* magazine, p. 26, with permission of the publisher. This article is copyrighted. © 1971 by CCM Professional Magazines, Inc. All rights reserved.

[15] Ibid., pp. 26–27.

CHAPTER 3

Getting Organized

The odds are that the reader's interest in the subject of business development is occasioned by one of the following three conditions:

1. Up to this point his firm has had no formalized approach to new client acquisition; perhaps he has become disenchanted with the results from chasing leads obtained from purely social contacts and he is looking for a businesslike approach to getting business.

2. He has had a halfhearted, basically hit-and-miss business development program which intensifies with the completion of current work, but tends to slack off when his staff gets busy. The peaks and valleys caused by this on-again, off-again pursuit of clients distresses both his associates and his accountant, not to mention his creditors.

3. His firm has had a reasonably successful business development program over the years and has enjoyed a moderate-to-good annual growth rate as a result. However, he cannot shake the feeling that his present approach needs a long, hard look, probably some pruning and touching up, and possibly even a complete overhaul. He would like to know what other firms are doing in this field and is receptive to picking up suggestions for improving his own efforts.

KNOW YOUR FIRM

Whichever category a firm falls under at this particular time, there are a few initial steps indicated before any meaningful decisions can be made about the intensity and complexity of its new or revised business development program. These steps may be loosely classified as internal investigations, under the heading "Know Your Firm."

1. Know your principals (partners and associates); your professional, technical, and office staffs, and their present and potential capabilities and specialties. It goes almost without saying that you should be equally aware of their drawbacks and limitations.
2. Know your general practice mix and—of equal importance—what is missing from it.
3. Establish realistic profit goals and attainable growth rates for the foreseeable future.
4. Know your competition—present and future.
5. Establish a desirable geographic coverage for your firm's operations. If expansion is indicated determine the best method of achieving it, e.g., purchase of other established firms or setting up your own branch offices; going into a new area "cold" or waiting until you have obtained a major job in the desired city, either on your own or with an associate.

Only after a genuinely objective and introspective look has been taken at a firm as it now exists and operates, coupled with an analysis of its goals for ten, fifteen, even twenty-five years from now, is one ready to embark upon an aggressive, effective, and professional program of business development. The short-range scatter-shot, hit-or-miss approach to client development is not nearly so productive in the long run as the reasoned, planned, and organized attack.

The principals may want to consider retaining an outside management consultant for all or part of this self investigation. Many professional firms do bring in outside consultants, with no particular relation to the design firm's size. The sad history of many such surveys is that the client heeds only what he believed or wanted to hear in the first place and ignores or denigrates the sections of the report which run counter to his preconceived ideas.

No matter whether the decision is to establish one's own investigative task force or hire a management consultant, the principals should be prepared to accept and act upon most, if not all, of the findings and recommendations for getting their own house and ideas in order.

MARKET RESEARCH

Other types of surveys may be helpful to future planning. Marketing studies, for example, can analyze client needs and thinking, qualify prospects, and measure the degree to which a firm is known by prospective clients and the general public.

In mid-1971 a large Midwestern firm of architects, engineers, and planners commissioned a marketing survey by mail among past and present clients scattered across the United States. The sponsoring architectural firm was not identified and the questionnaires were returned unsigned by the respondents. Among the results turned up by the study were these:

1. About two-thirds of the respondents indicated their organizations would require design and planning services in the next twelve months. (A 49-percent response was obtained from some 500 questionnaires sent out.)

2. Respondents were asked to rate seven considerations for selecting an architectural firm. The order in which the seven factors finished is interesting in that it seems to explode a myth or two about how and why clients select their architects. The seven factors and their ranking:

Total-service capability
Engineering know-how
Design creativity
Past projects of the firm
Postconstruction follow-up
Proximity of the firm to the project
National prestige of the firm

3. A question on client preference for the size (staff) of an architectural firm left the respondents almost evenly divided between "under 50," "51–99," "100 and over," and "No Answer." Each category accounted for approximately one-quarter of the replies.

In the summary of their report the market research firm that conducted the survey commented:

> . . . the respondents provided what we believe are valuable guidelines for new business activity. The survey confirmed that your strengths

are important to them and that your firm is better known than any other. It also provides a scale of importance that can be used to fine tune marketing activities and to guide you in adopting a posture that might be varied to fit the needs and desires of different prospects. One type of prospect, for instance, may prefer one "image" over another. And the large size of your firm is perhaps something to underplay in some cases. In other words, the results of this study may help you to "give them what they want."

After all, research is a marketing tool which should be used to supplement executive judgement; it's an advisory function that can be instrumental in business success if the data provided is useful to management.

THE DIRECTOR OF BUSINESS DEVELOPMENT

It is hoped that some of these studies and surveys will turn up one or more individuals among a firm's principals with an organizational bent. Regardless of how large or restricted the client acquisition efforts will be, one person must be assigned the primary responsibility for making it work. His job will be to coordinate all efforts, make assignments of principals and other staff where and when indicated, review promotion materials and presentations, maintain contact records, and follow up on all prospects. There should be a clear understanding with the rest of the staff members about their continuing obligations in obtaining new business, but every army requires a commander if it is to accomplish its mission.

It is probable, except in the largest of firms, that the person in charge of business development will be able to devote only part of his time to such efforts. For this and other obvious reasons it is vitally important that he be backed up in his work by an above-average secretary-assistant. Thoroughness, accuracy, conscientiousness, awareness, imagination, and alertness are some of the necessary attributes of this assistant.

A few of the larger practices have separate departments for business development, headed by a vice president or similarly titled executive. The staff, all full-time, of these departments varies from one to a dozen or more specialists.

RECRUITING A SALES STAFF

It might be instructive, at this stage of getting organized and regardless of the size of present practice or staff, to read a job and man description prepared a few years ago by an Eastern management recruiter in a search for a director of business development for an architectural/engineering firm. The document was purposely made vague in places by the recruiting firm to protect the identity of the client, who may be paying as much as $10,000 for the search efforts. In a situation such as this the client and the executive recruiter attempt to describe the ideal candidate, in the knowledge that some compromise will be required in the final selection.

BUSINESS DEVELOPMENT EXECUTIVE
FOR LEADING ARCHITECTURAL FIRM

THE FIRM: A highly respected and dynamic architectural firm. Its leadership is growth-minded and sales-oriented. Headquarters are located in a Middle Atlantic city, and operations are concentrated in, but not limited to the area within 500 miles of that city.

THE POSITION: The man selected for this position will be in charge of the business development function as well as a broad-gauged public relations program aimed at presenting the objectives, achievements, and performance of the firm to the financial, commercial, and educational public as well as to clients, potential clients, the public at large, and any other possible sources of commissions. He will represent the firm in new business negotiations, seek out prospects, cultivate them, and, with the assistance of key members of the staff, convert them into clients for the firm's architectural services.

Initial base compensation will be in the general area of $25,000, depending upon background and experience, plus liberal fringe benefits and opportunity for further professional and financial growth.

KEY FUNCTION: 1. Establish, in accordance with the firm's policies, new business development objectives and design plans and programs for their accomplishment.

2. Plan, in conjunction with others of the firm who have business development responsibilities, promotional programs having major prestige benefits to the firm. Such programs may concern the activities of the firm as a whole or be related to particularly outstanding projects.

3. Develop and maintain close personal relations with corporate, governmental, municipal, and other executives of client and potential client organizations, and otherwise keep properly informed on the timing of possible plans to utilize the firm's services.

4. Working independently, as well as with assigned prospects, present the firm's professional qualifications in such a way as to receive preferential consideration for a growing volume of projects that meet agreed-upon objectives as to diversification by type of client, type of structure, rate of profit, etc.

5. Establish and carry out a program of internal reporting, meetings, and communications as may be required in order to keep key

staff members and others with business development responsibilities
properly informed on prospects and negotiations with prospects in a
manner which will assure their fullest cooperation and support.

6. Initiate, encourage, and/or collaborate with others in the
preparation of articles for publication. Supervise the design and
maintenance of brochures or other material to document the firm's
capabilities.

7. In concert with key staff members, assemble and prepare
public relations communications regarding the firm's accomplishments, coordinating all such activity with external public relations
firms as may be required.

8. Establish cordial relationships with editors, publishers, and
news services dealing with the arts, architecture, and engineering,
as well as with other media executives who may assist in bringing
the firm's capabilities to the attention of those in related professions
and trades.

9. Disseminate to the general public through all proper and effective channels of communication (including the press, both public and
trade; publications; exhibits and expositions; reprints of articles and
speeches; etc.) information of general or professional interest, giving
particular attention to the timing of its release.

10. Plan and execute appropriate informational programs to specific groups such as: *(a)* professional societies; *(b)* trade associations;
(c) educational institutions; *(d)* industrial, legislative, governmental,
and civic groups; etc.

11. Assume responsibility for an internal public relations program
directed to employees, keeping them informed on all firm plans and
programs which may be of interest to them.

12. Represent the firm through membership and active participation in a wide variety of social, professional, and civic organizations
in such a manner that the firm's interest will be served to the best
advantage; stimulate other members of the organization to likewise
establish and maintain such public relations contacts.

13. Supervise the maintenance of files and records of all public
relations materials, including awards, clippings, and other documents
reflecting favorably upon the firm.

THE MAN: To receive consideration for this challenging assignment, a
candidate must demonstrate a combination of technical, administrative, and human relations skills of a high professional order. The
essential requirements are:

1. Proven background in the development and execution of strong,
imaginative public relations programs. A knowledge of the technical
and professional aspects of architectural practice is of course to be
preferred, but is not a mandatory requirement for the position.

2. Intimate, first-hand acquaintance with the technical aspects of
operating a public relations function (including the methods and
requirements of the various magazines and news media, procedures
for reproduction of graphic media, etc.).

3. Ability to program and coordinate his work on a practical day-to-day basis for the timely accomplishment of approved business development and public relations objectives.

4. Resourcefulness in utilizing publicity techniques so as to derive the greatest possible positive impact, within budget and staff limitations.

5. Creative human relations skills and the personal characteristics needed to develop and maintain, both externally and internally, those relationships of confidence and cooperation necessary to the fulfillment of his function.

6. Skillfulness in presenting his ideas effectively, both orally and in writing. Experience in speaking before groups and writing for publications would be helpful.

7. Aggressiveness in the pursuit of the firm's business development goals, yet the ability to work within a framework consistent with high ethical standards.

8. An intimate, first-hand knowledge of the major sources of architectural work. Also an understanding of current practices in real estate finance and leasing would be desirable.

9. Preferably, a degree in architecture (or engineering), although this is not an essential requirement for the position.

The man who was eventually hired for this position came from a similar post in another, smaller, East Coast architectural firm. His background and degrees were in public relations, rather than architecture or engineering (see number 9 under "The Man"). His starting base compensation was $28,000, plus an annual bonus amounting to between 15 and 20 percent of his base pay. These additional details are to indicate that successful business development specialists for architects, engineers, and related consultants enjoy a seller's market. A number of these positions in larger firms pay from $35,000 to $50,000 a year, plus various fringe benefits such as profit sharing and performance bonuses.

In this particular case, the first year's experience under the new business development director saw his firm's billings increase by $1.5 million. The previous year's growth in billings had been $500,000. While he would never claim full responsibility for the significant increase in billings over the year before, the new man must have done something right. Simple arithmetic shows that his first year's compensation of some $32,000 represented less than four-tenths of 1 percent of his firm's total billings of well over $8 million for that period.

Another, much briefer specification for a business development staff man was issued a few months ago. Prepared by another firm of management recruiters (executives search service) it read:

CONFIDENTIAL POSITION SPECIFICATION

JOB TITLE: Account Executive, Eastern Region
REPORTS TO: Manager, Eastern Office (dotted line)
                Director of Marketing (straight line)
COMPANY: Well-established architectural and engineering firm
                generating in excess of $200 million in annual con-
                struction and design commissions.·
EDUCATION: College degree in architecture preferred. Candidates
                with engineering degrees and strong interest in archi-
                tecture will also receive interested consideration.
AGE: 35 to 50 years—preferred to 45.
WORK EXPERIENCE AND ABILITIES:

1. Must have the personal qualities, professional selling skills, and level of empathy to effectively relate to
the needs of existing clients, identify and cultivate new
clients, and assume overall responsibility for business
development in the Eastern Region.
2. Should have a minimum of ten years experience
closely related to business development of architectural and engineering services or in the sale of products which require close contact with architectural
and engineering firms.
3. Must be a resourceful energetic self-starter capable
of functioning effectively with only a minimum
amount of work direction.
4. Must have the capacity for professional business
development to establish, maintain, and execute a
marketing plan to assure continued growth in the
Eastern Region.
5. Should have the administrative capacity to develop and prepare all periodic and special reports
desired by the Director of Marketing.

COMPENSATION: $20–25,000 base salary
                performance-oriented discretionary bonus
                profit-sharing plan

BUDGETING FOR MARKETING COSTS

Much as in the first example, the man selected for this job did not
have a degree in architecture or engineering. He did have heavy
experience in business development with other large design firms,

particularly in the East. His starting salary was almost $30,000 annually, plus a discretionary bonus based on his overall perform-ance.

With these two case studies of how firms staff up their marketing efforts, let's return to the subject of finances. Following the com-pletion of internal studies and the selection of someone to direct business development, the next step is to establish cost estimates and a budget for the marketing program.

Since promotion budgets of design professionals are seldom made public, one must interpolate, interpret, and adapt to come up with specifics on the subject. The American Management Association says an average figure for sales or marketing expense is around 10 percent, and includes the time of partners and other principals spent on new business activities.

In one recent seminar on marketing professional services the participants said their firms budgeted or spent on business develop-ment amounts ranging from 10 to 20 percent of annual billings. The largest figure was for a highly specialized consulting firm.

CORPORATE PROMOTION COSTS

The editors of *Advertising Age* annually devote most of one issue to the 100 U.S. corporations with the largest advertising and promo-tion budgets. Results for 1970, as carried in *Advertising Age* for August 30, 1971, showed that about one-third of the leaders had cut back on their advertising outlays as the economy shifted into a lower gear. While the decrease was less than 1 percent (from $4.64 billion to $4.62 billion) it was the first cutback recorded in the six-teen years the magazine has made the annual tabulation. In 1966, for example, spending on all types of promotion by the top 100 in-creased 12 percent over 1965. A few examples from the 1970 rank-ing might be of interest.[1]

Note that of the nine examples given, all but two of the companies (General Motors and Schlitz Brewing Co.) moved up in the 1970 ranking. Proctor & Gamble, as number 1, had nowhere to go, of course. The decrease in advertising and promotion spending in the recession year of 1970 shows that it is not only the small mer-chant who pulls in his advertising horns when the economy drops off. This cutback in such spending in bad times is in direct opposi-tion to the advice of many experts, but it seems to happen as regu-larly as day follows night.

Adver-tising Rank	Company	Advertising	Sales	Adver-tising as % of Sales
1	Proctor & Gamble Co.	$265,000,000	$3,178,081,000	8.3
2	General Foods Corp.	170,000,000	1,975,583,000	8.6
4	General Motors Corp.	129,764,000	18,752,354,000	0.7
10	AT&T	86,600,300	16,954,881,000	0.5
12	Coca Cola Co.	77,100,000	1,606,401,160	4.8
51	Standard Oil-N.J.	34,836,000	18,144,000,000	0.2
56	Schlitz Brewing Co.	30,750,000	594,437,000	5.2
78	Block Drug Co.	23,000,000	75,667,000	30.4
88	Eastern Airlines	19,000,000	971,050,000	2.0

The 1965 rankings and expenditures for these same companies, according to *Advertising Age* for January 2, 1967, were:[2]

1	Proctor & Gamble Co.	$245,000,000	$2,243,177,000	10.9
2	General Motors Corp.	173,000,000	20,733,982,000	0.8
3	General Foods Corp.	120,000,000	1,381,049,000	8.7
14	AT&T	69,900,000	11,323,000,000	0.6
16	Coca Cola Co.	64,000,000	518,424,872	12.3
36	Schlitz Brewing Co.	35,000,000	324,043,086	10.8
60	Standard Oil-N.J.	22,242,500	12,493,031,000	0.2
91	Block Drug Co.	15,000,000	36,000,000	41.7
96	Eastern Airlines	13,330,000	507,524,000	2.6

Even though the 1970 figures are now several years out of date, current ratios of advertising spending to sales probably are about the same. Note the 1970 variations in advertising outlays as a percentage of sales ranges from Standard Oil of New Jersey's 0.2 percent, on sales of over $18 billion, up to 30.4 percent for the Block Drug Company on sales of over $75.5 million.

PROMOTION BUDGETS FOR THE PROFESSIONAL

Obviously, one cannot directly compare the sales and advertising figures of a cosmetic company, for example, with those of an architectural or engineering firm. It would be flawed logic to attempt to relate $170 million in construction costs to $170 million in total sales of a home appliance manufacturer. In 1971 one of the country's leading design firms allocated 4 percent of its almost $13 million in billings for all forms of marketing and promotion. On projected billings of over $15 million in 1972, the firm had a promotion budget of something over $500,000 on the same 4-percent ratio.

As a base point for setting up a budget for sales efforts, business development, promotion—whatever the activity is called—a figure of between 5 and 8 percent of annual billings has been suggested. On $1 million in projected billings for the year, we have a tentative budget range of $50,000 to $80,000.

This is probably the most logical manner in which to arrive at a budget for business development. With the total amount in mind, the number of man-hours it will pay for and make available for new business efforts can be computed easily.

An alternate way is to decide in advance which staff members are to be assigned to business development, estimate the time each may spend, multiply this by man-hour costs, and thus arrive at the total figure. If the 5- to 8-percent guideline has been exceeded by this method a cutback may be in order.

The percentage-of-billings approach to budgeting for business development should be used, some marketing experts say, only as a starting point for calculations as to how much money would be allocated if conditions remained the same. "Dollars are sometimes allocated to promotion through emulating the competition and spending approximately the same amount. . . . This method assumes that all competitors have similar objectives and face the same tasks—a most dubious assumption."[3]

RECORD KEEPING

Let us assume that a director has been selected and a budget established for the business development program. Maintenance of contact records was mentioned earlier as one of the responsibilities of the director. Complete and accurate documentation of client acquisition activities is just as important as properly executed working drawings and specifications for any job turned out by an office for a client. An initial contact information form along the lines of the one shown in Figure 3-1 is used by some offices.

Other firms go in for a much more detailed form, such as the report shown in Figure 3-2. This format is set up to enable the information to be fed into a computer, but the information called for would furnish the basis for an excellent prospect file of any type.

Copies of the information sheets are distributed to all staff members who might reasonably be expected to learn of potential clients.

NEW BUSINESS WORKSHEET

Please answer as many of the following questions as possible and as fully as you can. If more space is needed for any subject use the back of this sheet. Be sure to note the number of the question continued over, for easy reference.

1. PROSPECT NAME _____

2. DATE OF INITIAL CONTACT _____ HOW MADE
   ☐ Telephone
   ☐ Letter
   ☐ Personal

3. CONTACT NAME(s) _____
   (Title)

   (Include, if possible, name of person who decides on architect)   (Title)

4. ADDRESS OF PROSPECT _____

5. CITY, STATE, ZIP_____

6. TELEPHONE NO. _____ EXT. _____

7. TYPE OF BUSINESS _____

8. PROJECT TYPE
   ☐ Office - High Rise   ☐ School   ☐ Research
   ☐ Office - Low Rise   ☐ University   ☐ Industrial
   ☐ Hospital   ☐ Church   ☐ Planning
   ☐ Governmental Bldg.   ☐ Airport   ☐ Other _____

9. APPROX. COST
   ☐ Under $1 million   ☐ $15-20 million   ☐ $50-75 million
   ☐ $1-5 million   ☐ $20-30 million   ☐ $75-100 million
   ☐ $5-10 million   ☐ $30-50 million   ☐ over $100 million _____
   ☐ $10-15 million

10. APPROX. SIZE _____
    Examples: Office Bldg.: no. of floors and/or gross sq. ft.; Hosp.: no. of beds

11. DATE CLIENT WANTS CONSTRUCTION TO BEGIN _____

12. OTHER ARCHITECTS BEING CONSIDERED (Names if possible)
    _____
    _____

13. DATE FORMAL INTERVIEWS SCHEDULED _____ WHERE _____

13a DATE ARCHITECTURAL SELECTION TO BE MADE _____

14. IF CLIENT IS CITY OR COUNTY GOVERNMENTAL UNIT, PREDOMINANT
    POLITICS (party) _____

15. INFORMATION SOURCE FOR THIS PROSPECT_____

16. OTHER DATA AND COMMENTS (use reverse side)

17. NAME OF PERSON MAKING THIS REPORT_____ DATE _____

Send original of this worksheet to Communications Department. Retain copy in your file.

CONFIDENTIAL

Fig. 3.1

TELEPHONE QUALIFICATION

The form in Figure 3-2 was set up to expedite telephone qualification of prospects by business development specialists. The six-step process of client acquisition, of which qualification is the second stage, is discussed in detail in the next chapter.

It is no secret that some people have a psychological block about seeking information or selling by telephone. AT&T has a ten-year-old program called "Phone Power," operated by more than 200 specialists throughout the United States. The Phone Power representatives hold 1-day seminars for small groups in six different programs:

1. Selling existing accounts.
2. Selling on the service call.
3. Reactivating old accounts.
4. Opening new accounts.
5. Collecting overdue accounts.
6. Qualifying prospects and making appointments.

Number 6, "Qualifying Prospects," has direct application to business development in design firms, but has been requested by a surprisingly small number of such firms. The training courses make use of self-instructional texts, along with lectures and practical application in the class.

A call to the local Bell Telephone affiliate (ask for the Phone Power specialist or the marketing department) will bring a Phone Power representative to your office to evaluate your needs and recommend a training course tailored to these needs. Single copies of the self-instruction books are available on request.

The staff member originating the client information form is asked to fill it in as completely as possible before routing it to the person in charge of new business. Note that many of the questions on the first form are designed for a check-off answer, while entries on the second illustration are, for the most part, simply circled. The business development office will be able gradually to fill in most of the remainder of the information requested. This particular form is not necessarily the best or the only approach to developing basic information about a potential client; you may want to adapt parts of it to your individual office requirements.

In addition to the initial information form, card files for quick reference and correspondence files for detailed coverage of pros-

PROMOTION DATA FILE - INPUT REPORT NO. 1

1 First contact record

2 Revise existing data   This date

Client (prospect) name

Street address

City   State (Province)   Country

First knowledge of prospect by (employee name)

1 Prospect initiated (previous client)

2 Prospect initiated (close acquaintance)

3 Prospect initiated (firm's reputation)   Remarks required

4 From publication

5 From rumor

6 Initiated cold

7 Other

This new prospective client status   Job number

1 Current client (non-related project)

2 Current client (related project)

3 Previous client (but not current)

4 New prospective client

5 Uncertain (why?)

This report by_____
                    (Name)

Fig. 3.2

```
┌─────────────────────────────────────────────────────────────────┐
│                                           (    )                  │
│  Primary client contact name              Phone number           │
│  ─────────────────────────────────────────────────────           │
│  Title or function                                                │
│                                                                   │
│  Project type                      Project financing (explain)    │
│                                                                   │
│                                    1   Unknown                    │
│  ──────────────────────────                                       │
│  Primary                           2   Excellent                  │
│                                                                   │
│                                    3   Apparently good            │
│  ──────────────────────────                                       │
│  Secondary                         4   Questionable               │
│                                                                   │
│                                        Explain ──────────         │
│                                                                   │
│                                    ──────────────────             │
│  Owner type                                                       │
│                                                                   │
│  1   Undefined                                                    │
│                                                                   │
│  2   Private/individual            Site selected?                 │
│                                                                   │
│  3   Private/corporation           1   Not applicable             │
│                                                                   │
│  4   Non-profit                    2   Selected (existing)        │
│                                                                   │
│  5   Federal government            3   Selected (new)             │
│                                                                   │
│  6   State government              4   One of several             │
│                                                                   │
│  7   Local government              5   To be found                │
│                                                                   │
│  8   Other                         6   Unknown                    │
│  ──────────────────────────                                       │
│  Explain other                                                    │
│                                                                   │
│  Separate new building?            We to assist site selection?   │
│                                                                   │
│  1   Not applicable                1   Yes                        │
│                                                                   │
│  2   New                           2   No                         │
│                                                                   │
│  3   Expansion                     3   Uncertain (Explain)        │
│                                                                   │
│  4   Renovation                    ──────────────────            │
│                                                                   │
│  5   Uncertain (Explain)           ──────────────────            │
│                                                                   │
│  ──────────────────────────       ──────────────────            │
│                                    (Name) Primary contact         │
└─────────────────────────────────────────────────────────────────┘
```

Fig. 3.2 (Cont.)

PROMOTION DATA FILE - INPUT REPORT NO. 2

1 First contact record

2 Revise existing data
_____
This date
_____
Client (prospect) name
_____
City   State (Province)   Country

Possible contract type   Is there a possible significant project
                                        dependent on or following this project?
1 Uncertain

2 Study/master plan   1 Yes   2 No   3 Uncertain

3 Full service   _____
                                        Possible project type
4 Thru sch. des.

5 Thru des. dev.   _____

6 Contract documents   _____
                                        Potential project dollar value
7 Other_____
_____   _____
Explain other   Pot. project fee if acquired (in $)

Association data   _____
                                        Primary associates name
Assoc?   Prime?
                                        _____
1 Yes   1 Yes   Secondary associates name

2 No   2 No   Prospect status

3 ?   3 ?   1 Speculative/not defined
_____
Explain ?   2 Defined, year + away

                                        3 Defined, less than year

                                        4 Invited to interview

                                        5 Interviewed (5 or more firms)

                                        6 Interviewed (2 to 4)

This report by _____
                    (Name)

Fig. 3.2 (Cont.)

Construction dollars defined

1 Uncertain

2 Probably less than ———————

3 Probably about ————————

4 Limit of ——————————

5 Not applicable

Probable construction $

$ ———————————————————

Fee basis

1 Not defined   4 D.P.E.

2 Fixed fee   5 Other

3 Percent

——————————— Percentage to our firm

Probable fee to our firm

$ ———————————————————

Prospect status (continued)

7 Proposal requested

8 Our firm only (or placed 1st)

9 Acquired, no contract

10 Acquired, contract

11 Lost - project folded

12 Lost - to another firm

1 Holding - good

2 Holding - doubtful

Explain ————————————————

————————————————————————

Have key decision makers been identified?

List key decision makers

Name

————————————————————————

————————————————————————

List names of known competing firms

————————————————————————

————————————————————————

1 Yes

2 No

3 Some

Title or function

————————————————————————

————————————————————————

Fig. 3.2 (Cont.)

PROMOTION DATA FILE - INPUT REPORT NO. 3

_____

This date
_____

Client (prospect) name
_____

City   State (Province)   Country

Significant events   Date

1 Letter of interest due   _____

2 Interview scheduled   _____

3 Proposal requested   _____

4 Expected award date   _____

5 Other (Explain) _____

Contacts since last update (use no more than 3)

1 Called   _____
                   Name

2 Received call   _____
                   Name

3 Wrote   _____
                   Name

4 Visited   _____
                   Name

5 Visited by   _____
                   Name

Significant update comments (as result of phone, letter or visit)

_____

_____

_____

_____

This report by _____
                (Name)

Fig. 3.2 (Cont.)

Reminders   Date   Who (use no more than 3)

1 Call by   _____   _____

2 Call expected   _____   _____

3 Write by   _____   _____

4 Letter expected   _____   _____

5 Visit by   _____   _____

6 Visit expected   _____   _____

General comments or thoughts concerning this prospect

_____

_____

_____

_____

_____

_____

_____

Fig. 3.2 (Cont.)

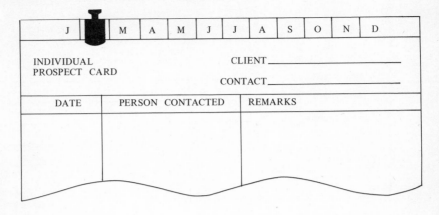

Fig. 3.3

pects and contacts should be daily reference tools for the people involved with business development. Examples of a card file system are shown in Figures 3-3 through 3-6.

This is a four-card system, covering prospective, present, and past clients. The first card is the Individual Prospect Card. This is begun as soon as the new business worksheet is received. The prospect's name and the firm's principal who will be the primary contact go on the lines in the top right corner. The twelve lettered divisions along the top of the prospect card are follow-up reminders. On the example shown in Figure 3-3 the Graffco metal signal tab is on "F" for February.

Every time a contact is made the date of the contact and name

PROSPECT MASTERCARD						
	CONTACT					
	CONTACT RECORD					
PROSPECT NAME	DATE	DATE	DATE	DATE	DATE	DATE

	CONTACT RECORD					
PROSPECT NAME	DATE	DATE	DATE	DATE	DATE	DATE

Fig. 3.4

of the person contacted are entered in the appropriate columns. Space is also provided for brief notes about the contact. These notations should be brief since a memo covering all important points of any meeting or other type of contact will be in the file folder for this prospect. Additional room is provided on the back of the card to continue contact notes. In the unlikely event that the prospect card is filled before the prospect is resolved, a second card is stapled to the front of the used-up card.

If it is determined that monthly contacts with a particular prospect are desirable, then the colored signal tab would be moved ahead a month at a time. From the prospect cards, following the

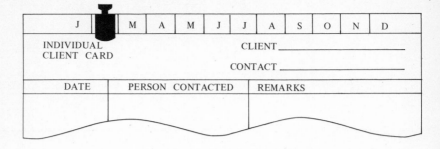

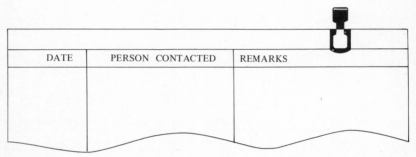

Fig. 3.5

tabs, the business development office prepares and distributes a list of contacts for each principal on a regular schedule; say the first Tuesday of each month.

The Prospect Mastercard (Figure 3-4) consolidates information about the prospects assigned to each principal. The principal's name goes on the "contact" line in the top right-hand corner of the front of the Prospect Mastercard. Every time an entry is made on the Individual Prospect Card a companion notation is made on the Mastercard. The Prospect Mastercards also provide a check for the business development office on how many prospects are assigned to an individual—and whether or not he is overloaded at any given time.

The third and fourth cards in this particular system are similar to the two already described, except that they are for following up with past clients. As soon as a job is completed an Individual Client Card is prepared in the business development office. Nor-

CLIENT MASTERCARD							
			CONTACT				
		CONTACT RECORD					
CLIENT NAME	DATE	DATE	DATE	DATE	DATE	DATE	

		CONTACT RECORD					
CLIENT NAME	DATE	DATE	DATE	DATE	DATE	DATE	

Fig. 3.6

mally the partner-in-charge or the project architect on the job would be assigned as the continuing contact. Note that there are twelve monthly divisions along the top of the Individual Client Card so that the signal tabs may be utilized as visible reminders to the business development office to notify the assigned contact in the firm to follow up with the client.

Some of the firms with access to computer facilities—either owned or leased—have been investigating the application of probability factors to their marketing efforts. These investigations usually take the form of predicting which prospects will become clients, at what fee and in which month, along with giving management a check on the individual effectiveness of the marketing staff members.

B.C. (before computers) some professionals approximated the computer printouts with a rating form for prospects along the lines of the form shown in Figure 3-7.   In that the marketing staff member had to make value judgments on the points to be assigned, the completed form usually was more subjective than objective, but it was at least a start on the rating of prospects.   On the other hand, it often became just one more piece of paper to shuffle through the internal communications system.

Some firms will carefully hoard every scrap of paper connected with the concept development of a project—but the principals resist the idea of writing even a short memo on their business development contacts.   Part of this, perhaps, is ego; the idea that one can keep all such multifarious details in his head.   Even when this is the rare case, unfortunately the remaining principals and others involved in business development are not usually gifted with ESP, so the intelligence is not transmitted.   The blanks so left in a prospect file can be very important to securing the commission.

On the subject of hoarding every line concerned with the early stages of schematics, this aside: The principal of one large firm, with a flair for show business effects, has a habit of sketching out ideas on the white plastic-covered top of the large conference room tables in his office.   After a lengthy client conference, the table top is often covered with sketches, scribbles, and annotations. Sometimes a client gets caught up in the spirit of the occasion and pencils in a few touches of his own, but that sort of participation is not encouraged in this particular office.

One evening, after business hours and following a design development meeting with a major client, the cleaning staff washed the table top clear as usual.   The next day one of the sketches on the table suddenly became very important to the job—and therefore to the architects.   No one could recall exactly how the principal had drawn it—including the artist—so an order was issued that all table tops were to be photographed after every client meeting. This procedure is still carefully followed and the file of table-top photos has grown to an alarming degree.

Some members of the firm have always thought that a little more "show-biz" mileage could be wrung from this theatrical

PROSPECT RATING FORM

Name _____

Address _____

	Multiplier	Rating Total
1. Selling distance from branch office ____mi. (0 mi = 10 pts. 1000 mi = 0 pts.)	0.5	_____
2. Production distance from main office ____mi. (0 mi = 10 pts. 2,000 mi = 0 pts.)	0.5	_____
3. Degree of political involvement (City Hall = 0 pts. None = 10 pts.)	1.0	_____
4. Size ($50,000 fee = 0 pts. More than $500,000 = 10 pts.)	3.0	_____
5. Timing (Indeterminate or long range = 0 pts. Immediate = 10 pts.)	3.0	_____
6. Future potential with client (No future = 0 pts. Campus type = 10 pts.)	2.0	_____
7. Future potential in market (No future = 0 pts. Expanding market = 10 pts.)	1.0	_____
8. Our firm's superiority (Churches = 0 pts. Commercial, governmental, medical = 10 pts.)	1.0	_____
9. Competition (Superior = 0 pts. Inexperienced local = 10 pts.)	2.0	_____
TOTAL POSSIBLE POINTS		140

_____
Name of rater

Fig. 3.7

technique if the photography crew would be brought in before the client left and while good byes were being said.   Most clients would have to be impressed with this evidence of thoroughness on the designer's part.

There is also the well-known story about Edward Durell Stone sketching the design for the facade of his U.S. Embassy in New Delhi on the back of an envelope during an airplane flight.   The envelope was discarded somewhere along the line—Ed Stone didn't really have to have it to remember his famous concrete grill and slender column concept—and carefully retrieved, the story goes, by one of his staff with an eye to architectural history.   In one of Stone's books there is a picture of the famous envelope, wrinkled and slightly coffee stained, with the Embassy sketch startlingly clear in spite of everything.   For a time certain nonbelievers caballed to spread the word that the envelope shown in the book was only a carefully prepared reconstruction of the original, which had gone the way of all airplane waste many years before.   Be this as it may, Ed Stone has gotten a lot of mileage out of the story.

STAFF INVOLVEMENT IN PROMOTION

The business development staff should not operate in a vacuum or be relegated to a "private club" status for a few top echelon principals and department heads.   Involve the entire staff in promotion planning and client acquisition discussions.   Bringing staff members into the picture through periodic meetings, all staff memos, and informal discussions gives them a feeling of being close to the center.   The interest and loyalty this open approach can engender will repay the principals many times over for the small investment of their time required to bring it off.   Everyone, regardless of title or responsibilities, likes to be on the inside.

It should be made clear at the outset that much of the information disseminated about client acquisition activities must be considered confidential and not for discussion outside the office. While there will always have to be some exceptions to open discussion of potential clients—the client may still be involved in assembling his site through a front, for example—it is a good idea to keep the staff posted in detail on the progress, or lack of it, in pursuing a particularly significant prospect, provided client confidences will not be violated.

MANAGEMENT INFORMATION CENTER

Some firms set up information centers to speed and spread the word about prospective clients among staff members. Sometimes called Management Information Centers (MIC), these information centrals often combine listings showing the status of current work with new business development. A simple form of an MIC would have one wall given over to a chart listing jobs going through the house, staff assigned, and present status (in schematics, under construction, etc.). Another wall would have a chart listing all potential clients, with columns showing the name of the client's contact, scope of the job, date of last contact, the firm's principal assigned to follow up, and similar information. (See Figure 3-8.) All business development files are normally kept in the MIC.

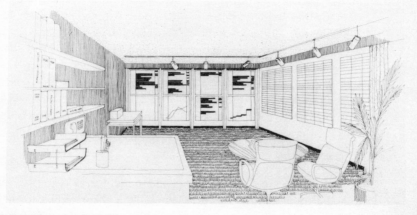

Fig. 3.8 *Typical management information center.*

One firm, with a sophisticated, well-equipped MIC, requires each of its principals to visit the MIC at least once a day, to insure their keeping current on all jobs—both in design and in potential. The partners are also expected to make a regular information input to the information center. An internal memo set out the types of information desired:

Current job problems and lessons learned.

Information about new or old prospects and past clients, including trips, presentations, meetings, or other contacts.

Reports of business or government conditions affecting practice.

Reports of current client presentations and reactions; upcoming groundbreakings, dedications, etc.

Any recent critical client contacts, such as a call from the president of ABC (a current client) advising that the chairman of the board of company XYZ (a potential client) will be visiting ABC's new building and asking us to be represented.

Information about our colleagues and competitors.

News of awards and other honors received.

AUDITIONS

A firm should identify several members of its staff who indicate they like the idea of participating in client acquisition activities. To insure that the full load of making presentations to potential clients and the important follow-up contacts are spread around among other principals and staff, some firms regularly hold what might be termed "auditions." These are dry-run, or mock, presentations, with members of the firm taking the part of the client. It is an unfortunate fact of life that all architects, engineers, and planners are not cut out to be outstanding public speakers—any more than every architect is destined to be a prizewinning designer. This subject is covered in detail in the chapters on presentations.

PLANNING FOR GROWTH

As to the practice mix—the second consideration in the internal investigation—it is surely obvious that a firm makes more profit from one $40 million project than it does from twenty $2 million jobs, everything else being equal.

One firm's principals decided that an average of $2.5 million in new work was required each week to maintain the rate of annual growth established over the previous five or six years. While all of the principals had generally been aware of the fact that a certain amount of new business was required each year to fuel their fairly large firm, no one had ever bothered to break it down to this precise weekly need. The general state of shock this fact induced initially almost convinced the accountant that his calculations had been a mistake—or at least the public announcement of the results.

He was able to ease the blow somewhat by pointing out that two or three $10 and $20 million jobs took care of a number of weeks, but for some period following the disclosure of the $2.5 million weekly requirement the managing partner was observed muttering "$2.5 million a week!" all the while shaking his head as in utter disbelief. In spite of this one accountant's experience, it is prudent and advisable for firms to have a similar check run on their own billings.

The arithmetic is not involved. One method of arriving at a weekly or yearly required input of new work for the coming year is to project the unserviced fees for the firm as of December 31 of the current year. Then estimate as accurately as possible how much of this amount will be earned in the next year. This will give the next year's earned income if no new work is obtained—a highly unlikely situation. Next, determine how much in new billings will be needed to just duplicate the present year's results and, finally, multiply by the hoped-for annual growth factor.

As an example, let's look at a firm which expects to have $1 million in unserviced fees at the end of this year. The principals estimate that 80 percent of the $1 million carry-over will be earned during next year—or $800,000. Billings for this year should total $2.5 million. The firm has enjoyed a 15-percent annual growth rate for the past seven years, so for next year, on this basis, the firm should achieve $2,875,000 in gross billings ($2.5 million plus 15 percent of the $2.5 million).

Since the firm will begin next year with the expectation of earning $800,000 in billings for work already in house, simple subtraction shows a requirement for $2,075,000 in new billings for next year. Keeping in mind that only part of this new work—say 75 percent—will actually be earned in the next twelve months, we determine that almost $2.8 million in new billings, in addition to the assured $800,000, will be needed next year to maintain the 15-percent growth rate. Using an average fee of 5 percent, the $2.8 million translates into some $55 million in new work required to earn the necessary billings. This can then be broken down into monthly or weekly goals for the business development office.

ANALYZING THE PRACTICE MIX

The practice mix analysis can turn up some surprising facts. One large architectural firm found that after twenty-five years of practice and some $2 billion in work its clients had been almost equally divided between public (52 percent) and private (48 percent) projects. This division was pure happenstance; no conscious effort or direction had ever been exerted to achieve such an even mix.

This same firm discovered during its investigations that over 70 percent of its work came from past clients.   This statistic was considered both good and bad.   Good, in that by implication clients were satisfied with its work; bad, in that new clients weren't being brought into the mix in sufficient numbers.   It goes without saying that a large backlog of repeat clients generally makes the job of business development much easier.   The inherent danger in such a situation is that as the client's personnel undergoes normal changes, the designer's contacts may be fired, retired, or transferred to distant branches.   Continuing contact and a lot of educated guessing as to who the rising younger stars in a corporation may be are always indicated, whether one job or thirty have been done for a particular client.

Another firm analyzed all its jobs to get an accurate picture of which areas it was strongest in.   The principals could then easily see where the blanks were in their practice—in both building types and size of projects.   Some of the blanks thus disclosed were of no interest to this firm but other blanks were considered important.   Business development activities were concentrated on opportunities to cover these "holes" in the practice mix.

Another method of analysis is to develop a chart by building types, comparing dollars, square footage, location of the project, and the percentage each building type bears to the total dollar volume of the firm.

Some years ago a well-known soils and foundation consulting firm decided it was taking far too long to become recognized as an expert on earth dams.   In spite of the fact that at least four of the staff were experts on dams, the firm had never been retained to design one.   Another factor which contributed to their general irritation at being ignored for this project type was an in-house study which showed dam work should be a highly profitable undertaking for the firm.

Their solution was to have the four staff experts write a book about dams.   The book, predictably, was a good one and soon became a sort of bible on dam design and construction.   The firm, now established as an authority on the subject, soon had a number of dams to design.

Still another firm decided at one point that the design of com-

munity colleges was a good field to get into. This was at a time
when the two-year schools were beginning to pop out of the landscape like mushrooms. Initial research consisted of an evaluation
of architectural firms already in the field, followed by in-depth
discussions with educators to determine their likes and dislikes
about their new college facilities. Materials were prepared and
staff trained to speak at educators' meetings. The campaign was
capped with a two-day seminar on the design of community colleges, to which all educators in the state were invited. The architects handled all meeting arrangements, including speakers and
meals, but the college officials paid their own expenses. Naturally,
most of the scheduled speakers were from the corps of experts
which had been developed within the architectural firm. The end
of this story is still somewhere in the future, since the firm continues to win community college commissions as a result of its
imagination and ingenuity.

LOST-JOBS ANALYSIS

Another potential input to the general study of a firm's practice is
a thorough analysis of the jobs it has lost in the last five or six
years — to whom and why. A simple three-column form can be used
for this analysis:

Job	Date	Reason lost, if known, and to whom
Center City Community College, Indianapolis	April 1969	Did not make final cut. XYZ & Associates selected. Had 3 completed CCs to our one.
Fairfax, Va. Airport	March 1970	Client thought we'd be too expensive. AB&C selected. B has brother on airport board.
National Insurance Co. office tower, Phoenix	August 1970	Survived cut to 3; lost out in final selection. TUV partners selected. Last job we did for NIC had problems — this didn't help.

Over a period of time certain patterns should emerge from this
study, especially if the lost client is inclined to be candid about
why a firm was not selected. Most are. An office may discover
that it is consistently losing out on the final cut or that one firm
seems always to pick up the job whenever the two firms are in
contention. Another possible finding from call-backs is that one

or more of the principals seem never to be able to bring home the job.

Do not try to establish a trend or pattern without sufficient information. It takes time to review the jobs lost, assuming that the necessary records have been kept to enable the study to be instituted. For the findings to be significant the survey should cover at least the previous three years.

PAROCHIALISM AT THE STATE LEVEL

As part of one such study a consultant reviewed a firm's experience with one of the New England states. The state selection board for designers, consisting of five members appointed by the governor, meets periodically to interview architects and engineers and to award design commissions. The executive secretary of the selection board is very proud of his mailing list of almost 600 design firms from all along the Eastern seaboard and this entire list is circularized whenever jobs are to be awarded. Architectural and engineering firms indicate their interest in projects for which they believe themselves to be qualified and the selection board picks several firms to interview for each job. The criteria, as published, sound impeccable:

BASIC CRITERIA FOR EVALUATION OF DESIGNERS

Relative weights to be established for each specific project
1. Professional Experience (Quality and Scope)
    a. Total building-design experience
    b. Public-building experience
    c. Specific experience on this type project
2. Quality of Design
    a. Reputation—general
    b. Reputation—on projects for State
    c. Ability to interpret—client's requirement
    d. Ability to balance—forms versus functions versus economy
3. Design Originality
    a. Ability to solve design problems with original ideas
    b. Fresh approach
    c. Efficiently
    d. Economically
4. Cost Performance
5. Schedule Performance
    a. Plans and specifications
    b. Construction phase

6. Supervision of Construction
   a. Proximity to agency and job site
   b. Adequacy of supervising architect
7. Available Resources
   a. Financial status
   b. Use of engineering specialists
   c. Staffing plans
8. Reputation Requirements of Project
9. Effective Relationship with:
   a. Client
   b. Bureau of Building Construction
   c. Contractors
10. Economic Benefit to the State
11. Other Special Criteria

The consultant's client had often been selected for interview but had never received a design commission. The designer selection board publishes an annual report in which the firms selected for state work are listed, so the reports for the last five years were carefully reviewed. This check turned up the discouraging fact that out of some 100 major projects, only one had gone to an out-of-state firm. Rather obviously, criterion number ten was weighted heavily in all selections. One cannot really argue with the idea of a state wanting to keep its funds at home, but for state officials to stage such an elaborate charade to give the impression of non-parochialism was a dirty and expensive trick on all of the out-of-state firms who made the annual pilgrimage to the selection board interviews in good faith.

UNIQUE PROJECTS

On the subject of diversification of a firm's practice mix, almost every architect, engineer, and planner has a favorite story about his most unusual project; one-of-a-kind designs which come along occasionally to amuse, baffle, or confound. For example, a Utah-based firm may be the only one in modern times which counts a house of prostitution among its recent completed works. Such institutions of pleasure apparently are legal in Elko, Nevada, and the firm's design was successful to the point that the other houses immediately began losing trade to the new facility. The architects have been approached by five other Elko madams interested in new quarters. While they feel they have stumbled on a valid speciality, no one in the firm is particularly anxious to capitalize on it. (The topic "Image" is discussed in detail in the chapter on public relations.)

Perhaps equally unique was the client who called one day to ask a designer if he would be interested in preparing plans for a mushroom farm. Since his call came during a rather slow period, jobwise, he was assured of the firm's interest, and a date was set up for a presentation. About all that was really known about the prospective job was that the potential client had seen a mushroom farm in Canada which he liked and his farm was to cover five to six acres of ground. It will come as no surprise to readers of this book who have designed mushroom farms that the structure resembled a large, low, windowless warehouse, with provisions for fairly strict temperature and humidity control.

Hangars for blimps and Boeing 747s, most of the installations at Cape Canaveral and in NASA's Houston space center, and certain other governmental projects for sophisticated research would all fall under one-of-a-kind or specialty work.

An official of one of the Federal Reserve Banks tells the story of some unusual problems he once had with an architect. The bank required a large vault for long-term storage of certain coins and paper money, far removed from the city. A site was located back in some nearby mountains and a local architect was retained to draw up plans for the vault. One of the most important design considerations was that the entrance be all but invisible to passers-by. The location of the vault was known to only a half-dozen top officials of the bank, one armoured car team which would make the pick ups and deliveries—and the architect and contractor, of course.

Since so much thought and discussion had been devoted to the security aspects of the vault's location and construction, the bank's president was understandably surprised—and quite annoyed—to have the architect call on him one day with text, plot plans, and pictures of the completed vault, prepared for publication in one of the national architectural magazines. Yes, of course the architect understood the security considerations, but since this was a most unusual structure he knew it would be good for his reputation to have it published. The job was not published, needless to say, and that architect did no more work for any bank.

ESTABLISHING GEOGRAPHIC COVERAGE

The last point in our sequence of self-investigation steps is to establish a practical and desirable geographic coverage for a firm's operations. As a practical matter, a small office is usually limited to a regional practice—say a 100- to 150-mile radius of its home city. Larger firms may aspire to seminational (east of the Mississippi River, for example), national, and even international practices. The chapter "Overseas Client Acquisition" discusses the latter, while the chapter "Joint Ventures, Associations, and Other Consortia" explores nonbranch office methods of extending one's sphere of influence and practice coverage.

BRANCH OFFICES

The establishment or acquisition of branch offices enables a firm to spread its coverage on an organized and selective basis into areas which appear to offer the most promise of new business. Practically every design firm which elects to go the branch office route sooner or later discovers the importance of developing good intelligence about the new area. Some of the sources of such intelligence are:

1. Friends (preferably fellow architects or engineers in whose judgment one has confidence) who live in the communties under study.
2. Chambers of Commerce. Most chambers publish regular reports on their city's leading economic indicators: construction activity, bank deposits, employment, etc.
3. Past and present clients with branches or affiliates in the city or cities being investigated.
4. Newspapers and other local publications. Take out a three-month mail subscription to the leading newspaper.
5. Suppliers who may have branches in the area under study or salesmen who call on prospects and clients in the area.
6. Personal visits over a period of several months to the city under consideration by one or more of a firm's principals.

The importance of evaluating the competition in the new city at the same time should be obvious.

Branch offices are often opened concurrent with the winning of a large design commission in a new city, particularly if a firm has already investigated the area with expansion in mind. This is a relatively painless way of getting a branch office underway if it is planned to service the job locally. Sometimes a client will make the establishment of a local office a condition of the contract award, so the decision about opening a branch office essentially is made for the architect or engineer if he wants the job. We are not referring to field offices in this context, but branch offices have occasionally sprung from field operations. Experience with a field office in the area is always something to add to the total input when considering branch office locations.

Another approach is cold-bloodedly to set aside X dollars for start-up and operation costs of a branch office in a new area for a specified period of time: at least twelve months. At the end of the trial period the entire experience is carefully evaluated and a decision made to close it up or continue in operation. If enough work has come in to cover at least the original investment, the branch usually may be considered a success and the decision is not a difficult one. On the other hand, several prospective clients may be close to a decision and a careful analysis of the branch office's future prospects is called for by the parent firm.

IDEAL BRANCH OFFICE LOCATION

There is no question about branch offices helping to sell a designer's services in their areas. Many governmental clients require local representation. Branch offices, in the right location, can serve also as a reservoir of good new staff members, particularly for firms headquartered in large cities where top designers and qualified technical personnel are not readily available.

Established areas of population concentration can be an aid to deciding where and where not to locate branch offices. The major megalopolises have been pretty well defined by the Census Bureau: Boston–New York–Philadelphia; Washington, D.C.–Baltimore–Philadelphia; Cleveland–Detroit–Chicago; and Indianapolis–St. Louis–Chicago; but there are many others not quite so large or well known. The ideal location would be a fast-growing community with excellent transportation facilities; a young to middle-aged population; an outstanding climate; nearby physical features conducive to year-round recreational activities; no large architectural, engineering, or planning firms (or any of these firms of any size with established national reputations); and a good college turning out architectural and engineering graduates. Whoever finds such a place may set up a branch office with confidence in its future. On second thought, he should move his main office there and leave a branch office in his present location.

REFERENCES

[1] *Advertising Age*, Crain Communications, Inc., Chicago, Aug. 30, 1971, pp. 2, 29.

[2] Ibid., Jan. 2, 1967, p. 31.

[3] James F. Engel, Hugh G. Wales, and Martin R. Warshaw, *Promotional Strategy*, Richard D. Irwin, Inc., Homewood, Ill., 1967, p. 125.

CHAPTER 4

How and Where to Find Prospects

"Ethical purity, triply distilled, would restrain [the architect] from any more direct or semi-direct business chasing than is practiced by the family doctor. That would be exactly none at all, and so it was in the old days to a considerable degree."[1]

So wrote Royal Barry Wills some thirty years ago. "Business chasing" is now known—more politely if perhaps less accurately—as "business development," and is practiced in some form or another by every American design firm. The primary differences to be noted among offices result from the degree, organization, and success of the chase. As background for this chapter you might re-read Wills's list of twenty-three ways to develop new business in Chapter 1.

PACAT SYSTEM

The approach taken by some design firms to business development might be reduced to a Madison Avenue-sounding acronym: the

PACAT System.   PACAT translates as Participation, Alertness, Curiosity, Awareness, and Timing.

1. Active *participation* in community affairs.

2. *Alertness* to job leads from friends, family, and acquaintances in business, education, and government.

3. *Curiosity* to become *aware* of other disciplines and expertise beyond the normally assumed scope of knowledge—methods of construction financing, speculative building projects, and governmental participation in funding, to name a few.

4. *Timing* as applied to everything we do (the fourth and most critical dimension).

Another method of outlining the process of client acquisition:

1. Identification
2. Qualification
3. Investigation
4. Planning (strategy)
5. Pursuit
6. Close

PROSPECT IDENTIFICATION

Identification of prospects is one of the two most important keys to the whole process; the second is qualification, which we will discuss later.   One readily available and much-used source of prospect identification is *Fortune* magazine's annual May issue listing of the top U.S. companies.   This one source provides a list of 750 potential clients for a starter, covering the

500 largest corporations
50 largest commercial banks
50 largest life insurance companies
50 largest retail organizations
50 largest transportation companies
50 largest utilities

These 750 firms obviously will control or spend the lion's share of the corporate construction dollar.   Since these same companies, almost without exception, are investor owned, copies of their most recent annual reports may be obtained through their public relations departments or a friendly stockbroker.   A section of the annual report usually is devoted to the company's future capital expenditures—and is often accompanied by a general description of locations, building types, and scope.

Forbes' annual Directory issue, "Dimensions of American Business" (also issued in May), is at least as good a source of general information about large corporate prospects as that published by the *Fortune* editors.   *Forbes* measures four dimensions of the 500 largest American companies: revenues, net profits, total assets, and total stock market value.   The "Company Roster" section of *Forbes'* Directory issue includes the name of each company's chief executive, his salary, and the total remuneration of all officers and directors.

Several years ago a well-known firm of soils and foundation consultants developed a sophisticated computerized approach to identification of prospects for its rather specialized interests in business development.   Project types covered in this information retrieval system included:

Airfields and Airfield Facilities	Industrial Processing Plants
Bridges and Elevated Transportation Structures	Land Developments
	Marine and Submarine Structures
Buildings—to Twenty Stories	
Buildings—over Twenty Stories	Missile and Space Facilities
Canals	Pipelines and Penstocks
Communication Systems	Power Plants
Cryogenic Structures	Power Transmission Systems
Dams, Earth or Rock	Recreation Structures
Dams, other than Earth	Reservoirs
Docks and Marinas	Sewage and Water Treatment Plants
Drainage and Dewatering Facilities	
	Storage Elevators
Earth-retaining Structures	Tunnels
Groundwater Development	Wells
Highways, Streets, and Pavements	Underground Structures
	Others

Three-digit code numbers are assigned to each project type.   A second code master of the same firm's system covers the possible varieties of owner (client) types and subtypes:

Architect
Architect-Engineer
Contractor
Engineer
Engineer-Constructor
Financial, Insurance, Real Estate, Developers, Attorneys
Foundations and Institutes
Government
  Foreign
  Federal

Agency for International Development — AID
Air Force
Army
Atomic Energy Commission — AEC
Coast Guard
Department of Agriculture
Department of Commerce
Federal Aviation Agency — FAA
Federal Housing Administration — FHA
General Accounting Office — GAO
General Services Administration — GSA
Housing and Urban Development — HUD
Import-Export Bank

Inter-American Development Bank — IADB
National Aeronautics and Space Administration — NASA
National Park Service
Navy
Postal Service
Soil Conservation Service
United States Bureau of Land Management
United States Bureau of Public Roads
United States Bureau of Reclamation
United States Coast and Geodetic Survey
United States Forest Service
United States Geological Survey
Veterans Administration — VA
World Bank

  State
  County
  City
  Special Government Agencies
  United Nations
Industrial
  Agricultural, Farming
  Motor Vehicles and Parts
  Chemical
  Electronics, Appliances
  Food and Beverage
  Furniture
  Lumber, Wood Products
  Metals and Metal Products
  Mining
  Paper

Petroleum
Pharmaceutical, Drugs
Textiles
Construction Materials
Aircraft and Parts
Machinery (Office, Industrial)
Measuring, Scientific, and Photographic Equipment
Publishing and Printing
Rubber

Shipbuilding and Railroad   Soaps, Cosmetics
    Equipment   Tobacco
Religious
Utility
    Communication
    Gas
    Power and Light
    Transportation
    Waste Disposal
    Water Supply
Geologists and Geologists-Hydrologists
Commercial (Stores)
Home Owner/Individual
Educational
Hotels, Motels, Apartments
Others

Another standard breakdown of building types into six major divisions and eighteen subgroups is used by *Progressive Architecture* in its annual business survey. Initially, this less-detailed breakdown might be preferred over one similar to the previous, lengthy, building-type list:

Commercial
    Low-rise
    High-rise
    Industrial
Community
    Planning and design, nongovernment
    Urban design and redevelopment, including public housing
Federal government
    Office and service
    Hospitals/health
    Defense and space
    Other (not including housing)
State and local government
    Office and service
    Educational
    Hospitals/health
    Other (not including housing)
Institutional
    Educational, private
    Hospitals/health, private
Residential
    Private, single
    Low-rise (not including public housing)
    High-rise (not including public housing)

PROSPECT QUALIFICATION

Qualification, the second stage in the client acquisition process, deals with the important question: Is it a real job? In other words, is the project apt to go just so far down the line and then founder for one reason or another, or does it stand a reasonable chance of being built? Today's designer has little interest in designing vast projects—or even small ones—which do not go on to completion. Certainly, if he negotiated his contract properly, he will be paid for his work up to the point it goes on the shelf—but money and reputations are not made from plans gathering dust in some corporation's storeroom.

Two indicators of the credibility of a prospective job are the status of its financing and the stage of its program development. With some jobs these are very clear and identifiable signs of its potential, though usually requiring close attention to the media, including newspapers, financial and trade publications, and news magazines. Some of the tip-off signals are the creation of authorities for hospitals, ports, highways, bridges, and airports; public meetings to discuss the need for master plan or feasibility studies—for cities, hospitals, airports, and new towns, to mention a few possibilities; bond elections for schools, hospitals, and airports; reports of large-scale land acquisition by corporations or governmental units; and announcements of real estate zoning variance hearings. Occasionally, the job has already been awarded by the time these considerations become public knowledge, but it always pays to check it out.

THE CHECKLIST QUALIFIER

Some firms develop checklist questionnaires for several building types, e.g., hospitals, office buildings, and schools, so that in an emergency almost any member of the staff can go through the list with a potential client. This gets most of the necessary preliminary information in a standard format and, at the same time, keeps the staff person from sounding like an uninformed idiot if he is dealing with an unfamiliar building type. To many clients a building is a building—and they assume that any architect will automatically know all about every kind of structure and program. Seldom is this the case.

A Hospital Prospect Checklist for the initial (qualifying) contact might look like this:

GENERAL

1. Will the project involve replacement, remodeling, an addition, new construction, or some combination of these?
2. If replacement, remodel, or addition, when was the original building completed? Initial cost? Size? Are plans available?
3. Is there a master plan? How old is it?
4. Is the present facility accredited? If not, what is accreditation status?
5. How many beds? What type?
6. Are the present beds classified as conforming?
7. Size of the site?
8. Status and condition of present outpatient and emergency facilities?
9. Occupancy rate? Average stay?
10. What are the daily rates?
11. What is the fire classification?
12. For long-term facilities, how many on the waiting list?
13. Are the support facilities operating at capacity, or will they handle extra beds?
14. Overall condition and efficiency of present plant?

MEDICAL AND NURSING

1. Staff size?
2. Size of nursing stations?
3. What in-service training programs are offered?
4. Is teaching involved? If so, where do the students come from?
5. What percentage of patients are taken care of by the house staff? What percentage by private physicians?

FUNDING

1. Has the state certification of need (or its equivalent) been obtained?
2. If Hill-Burton participation is anticipated, has a letter of intent gone to the state comprehensive planning agency?
3. Who is the director of the state hospital planning agency? Where is his office?
4. Are there local or state limitations on design fees?
5. What funds are now on hand to enable planning to proceed?

SCHEDULE

1. Has an A-E selection committee been named? If so, the names of the committee?
2. When will formal interviews begin? How many firms will be interviewed?

3. Names of members of the hospital board or other operating authority?
4. When will the A-E be retained?
5. When is occupancy required (or desired)?

It is always preferable, of course, to avoid giving any kind of cost estimates in the initial, fact-finding meeting, but sometimes it just cannot be avoided. In discussing costs always talk in ranges and for rough estimating the following figures might be used.

For an all-new 100-bed hospital to include full services, figure on 850 square feet to a bed, or a total of 85,000 square feet. One might use a cost range of $35 to $45 per square foot. In this example that works out to between $3 and $3.9 million (estimated construction costs for a 100-bed general hospital facility). If the project is a beds-only addition, with a minimum of new support facilities, a lower per-square-foot figure of around $30 a square foot may be justified. Relate any estimates to the local market and conditions as best you can in these early stages.

The remaining four items in the client acquisition list—investigation, planning, pursuit, and close—will be covered in detail in the chapters on presentations. A comprehensive listing and evaluation of specific sources of prospect information, along with a number of recommended reference sources for the business development library, will be found at the end of this chapter.

PROSPECT SOURCES

Consultants retained by a firm should be an excellent external source of leads to new business prospects. Since most consultants work with several firms, one cannot expect an active structural engineering firm, for example, to throw every potential job into just one design office—but the relationship should be such that any one firm receives its share of tips on job leads.

If you have not enjoyed this kind of consultant input to your business development efforts in the past it is suggested that you draw up a list now of all the consultants used in the past three to five years. Beginning with the ones worked with most often, make it clear to their principals that you will appreciate (and expect) to hear regularly from them about job opportunities as they come across them.

The same point is true with construction firms. The principals and the head of business development for the larger constructors often hear about prospective jobs long before anyone else in the field. Since the designer is frequently in the position of being asked to recommend both consultants and contractors to clients, the process has benefits for all concerned. It has some obvious disadvantages for those who do not cooperate.

Suppliers and manufacturers' representatives can be equally productive of job leads. For example, one firm learned about a multimillion dollar school job from a furniture salesman who happened to mention it to one of the interior design department staff members. The client turned out to be one for whom the firm had done previous work so it was not difficult to obtain the commission—particularly since one of the principals was able to contact the school board long before any other firm knew about the project. That salesman guaranteed his future welcome in that office and, whenever appropriate, sales of his line for the firm's jobs.

STAFF PARTICIPATION AND RECOGNITION

Related to the example above is the question of participation of a firm's own staff in business development, as was touched on in Chapter 3. Every staff member has his or her own circle of friends and contacts and should be encouraged to develop job leads and business acquisition intelligence whenever and wherever possible. Good leads may be reported by staff members who know when to listen and what to listen for at cocktail parties and other social events. Not long ago a secretary was able to confirm several important points for her employer about the principals and sources of financing for a large project in New York City, just by happening to be seated in a group at a party where the project was being discussed.

This might be as good a place as any to discuss the often sticky problem of recognizing such staff contributions to job procurement. Perhaps even more difficult of solution is the question of how to handle upper-level staff people who are especially active in business development—particularly when they are developing specific leads and calling on potential clients. We are not talking about partners or associates in this regard; it is assumed that their activities in this field are both expected and motivated by hopes of building up greater profits in which to share at the end of the fiscal year.

Frankly, a truly equitable method of recognizing staff participation in business development is yet to be worked out. Some firms offer a percentage of the fee—up to 10 percent. It usually is difficult, if not impossible, to single out one individual as the sole person responsible for obtaining a specific job. Assume, then, that it is possible to attribute the obtaining of a job to three or four staff members, each of whom theoretically is entitled to a share of the allotted percentage of the firm's gross fee. Who then determines the extent of each man's or woman's participation, as a fraction of the total effort required? Few King Solomons are found in architectural and engineering offices.

Since recognition and reward are basic motivators for most people, one cannot just ignore the problem. Smart directors of business development make it a point to acknowledge and document all staff assistance and participation by memo to the managing partner or other principals, with copies to the personnel director and the individuals involved. The premise here is that management will give weight to these communications when promotions and bonuses are under consideration. It might be mentioned in this connection that during slack economic periods, such as the recession of 1970–71, the motivation tends to take care of itself. When staff members know of friends and fellow professionals who are out of a job because enough new work has not come into their firms, they become vitally interested in all aspects of job procurement, without an instant concern for monetary or other types of rewards.

ORGANIZATION ACTIVITIES

The advantages of membership and active involvement by the principals of a firm in a variety of civic, business, religious, cultural, political, and professional organizations should be evident. In addition, the same kind of participation should be encouraged on the part of associates and other senior staff people.

A half-dozen of architect-author Wills's suggestions for develop-

ing new business (Chapter 1) are pertinent to the subject of organization activities:

1. Talks before clubs
2. Civic service in town, city planning, or art commission
3. Recommendations from friends or club and fraternal acquaintances
4. Having a wife in club and civic work, who keeps her ear to the
ground
5. Making the most of college alumni associations
6. Regulated social activity, not wasted in the company of
other design professionals

If it has not been done recently, a good first step in assessing a
firm's coverage of local, regional, and national organizations is to
make an inventory of memberships currently held by all members
of the staff. With the inventory at hand, obtain the remaining
coverage desired through volunteers or assignment. Along with
the organization checkup an equitable policy for payment of dues
and fees should be implemented.
Some of the organizations which might be included in a firm's
coverage are:

Service clubs
    Rotary, Lions, Kiwanis, Exchange, Toastmasters
Business organizations
    Chambers of Commerce (local, state, and national), Jaycees,
    American Management Association, Sales and Marketing
    Executives International
Professional organizations
    AIA, NCPE, AIP, CEC, AID, American Hospital Association,
    American Association of School Administrators
College and university alumni groups
    Including membership on Boards of Trustees and Boards
    of Visitors

Religious, cultural and political organizations are too numerous to
list in this brief recap, but their importance should be obvious.
Professionals also may gain important exposure to their many
important publics through service on local, state, regional, and

national governmental boards and commissions, as well as by serving as a visiting professor or guest lecturer in their specialties at schools and universities.

OTHER JOB LEAD SOURCES

Still another source of job leads are fellow professionals. These are usually potential associates or joint venturers who, for various reasons, do not feel they can obtain or produce the job by themselves. Perhaps their firm is too small or too busy, or the job is too complex or the potential client has specifically asked for a certain kind of expertise they do not have in house. Whatever the reason, most firms are approached from time to time about entering into an association for a specific job. Chapter 11, "Joint Ventures, Associations, and Other Consortia," discusses this procedure in detail.

Contacts in fund-raising firms can be productive of early knowledge about prospects in hospital, association, and education fields.

Likewise, the cultivation of staff members of state development commissions is to be encouraged. These are state-funded groups whose assignment is to locate new headquarters buildings and industrial plants in their states. Even in a small state such as Connecticut an active development commission may be responsible for thirty to fifty major moves—and buildings—annually. The same is true of the development offices of utilities: railroads, power companies, and port authorities.

The present role of insurance companies, along with banks and certain union trust funds, as sources of financing for major projects of all types should be so well known as to require no further coverage here. The importance of contacts within these organizations is self-evident.

Air travel is a way of life for many design professionals today — but only an alert few effectively utilize their hours aloft for a type of captive audience presentation to their fellow travelers from corporations and government. Any number of commissions have had their genesis 35,000 feet up in the air, over a Scotch and water or a postdinner cup of coffee. Circumstances could be termed almost ideal for low-key soft selling: the relaxed atmosphere, an hour or two to while away, and no interruptions except by a pretty hostess looking after your personal comfort. Some design professionals take great pride in the fact that they travel several hundred thousand miles a year by airplane, but they have never brought a job down from the clouds. Others—the real salesmen— regularly find solid prospects on even the Newark to Boston air shuttle.

The most obvious source of new business has been left until last —past clients. Few governmental or corporate clients require only a single building program; i.e., they are continually building something somewhere. How much of their repeat work a firm falls heir to is pretty much up to its principals. The primary requirement for getting additional work from a client is to satisfy him the first time around. If that was not accomplished—and it happens to the finest of firms with the highest of intentions—best forget him and look elsewhere for job leads.

In government work, particularly at the federal level, usually there is the consideration of passing the work around. Be there and ready when your turn comes up. Turn-taking normally is not a concern of private clients, assuming past relations have been good. There is an inherent advantage to the corporate client in utilizing the same design firm again and again, in that the individuals involved become used to working with each other, saving time and—hopefully—money for the client. The main disadvantage of this system, from the corporate client's standpoint, is that he may end up with an unwanted sameness of approach and design in all of his projects. If both designer and client are on their toes and communicating, this should not become a problem.

SPECIFIC JOB INFORMATION SOURCES

Commerce Business Daily (Catalog No. C41.9): Published daily by the U.S. Department of Commerce. Available from the nearest Department of Commerce Field Office or the Superintendent of Documents, Government Printing Office, Washington, D.C. 20402. Price $40 a year; by air mail $70.25 a year.

Contains unclassified requests for bids and proposals, procurements reserved for small business, Postal Service construction-leasing proposals, contractors seeking subcontract assistance, upcoming sales of government property, prime contracts awarded, research and development leads, current foreign government procurement offers in the United States, and nongovernmental export opportunities for American firms.

Herb Ireland's Commercial Expansion Reporter: Published monthly in fourteen U.S. regional editions by Prospector Research Services, Inc., 751 Main Street, Waltham (Boston), Massachusetts 02154. A Canadian report is also available. Price for one regional edition is $42 a year; all fourteen U.S. editions, covering the fifty states, $295 a year.

Reports on construction plans for shopping centers, office buildings, stores, hotels, motels, banks, hospitals, nursing homes, apartment houses, recreational facilities, sports and athletic centers, college and university buildings, civic centers, and other municipal buildings and residential developments. Many of the projects listed in the *Commercial Expansion Reporter* are already in design and therefore of more interest to contractors and suppliers than to architects, engineers, and planners.

A companion service, *Herb Ireland's Sales Prospector,* reports on industrial expansion, new distribution centers, transportation terminals, and power-generating facilities in the same fifteen regional edition format. Subscription prices for the *Prospector* are the same as for the *Reporter.*

Engineering News-Record ("Pulse Section"): Published weekly by McGraw-Hill, 1221 Avenue of the Americas, New York, N.Y. 10020. Price $10 a year.

The "Pulse Section" carries the week's significant projects in planning and bidding stages. Occasionally a job is listed before the A-E has been selected.

Foreign Projects Newsletter: Published biweekly by Richards, Lawrence & Co., P.O. Box 2311, Van Nuys, California 91404. Price $150 a year. Reports private and governmental projects on a worldwide basis, except for continental United States. See the description of this newsletter in Chapter 12, "Overseas Client Acquisition," for more details.

National Building News Service, Inc.: NBNS, P.O. Box 647, Ridgewood, New Jersey 07451, publishes four job information letters:

NBN Weekly Report: Covers all types of buildings in the planning and bidding stages over $1 million in the United States and

Canada. Each weekly issue is in three sections: (1) new projects —first reports, (2) new data on earlier reports, and (3) bidding information.   One year $625; three years $1,575.

Medical Facilities Planning Report: Monthly listing of hospitals, nursing homes, and other proposed medical buildings at the "architect selected" stage.   Price $200 a year.

Educational Facilities Planning Report: Monthly listing of elementary, secondary, and higher education buildings at the "architect selected" stage.   Price $275 a year.   Both the Medical and Educational Facilities reports for $400 a year.

NBN Architects' Monthly Building Report: Contains listings of projects in Canada and the United States over $1 million for which no architect or engineer has been retained.   Projects listed in the monthly report are compiled from the *NBN Weekly Report.* Price $75 a year.

Service World International: Published by *Institutions Magazine,* 205 East 42d Street, New York, N.Y. 10017.   Free.

Service World International, upon request to its publisher, offers a reader service of providing leads on firms planning new construction or expansion.   SWI also maintains worldwide listings of distributors and agents selling to the hotel and allied industries.

b.i.d.s. Jobletter: Published monthly by Building Industry Development Services, 1914 Sunderland Place, N.W., Washington, D.C. 20036.   Price $100 a year.

The *Jobletter* covers all fifty states, both public and private work, with emphasis on federal projects.   Has occasional foreign prospect listings.   In addition to in-depth articles about Federal departments and agencies, *b.i.d.s* contains additional useful information about business trends, professional services marketing techniques, and other aspects of business development.   Special supplements and reports are available to subscribers on request.

Million Dollar Project Planned List: Published monthly by Live Leads Corporation, 369 Lexington Avenue, New York, N.Y. 10017. Price $540 a year.

Lists only projects of more than 50,000 square feet and/or $1 million in cost in the planning stage.   Projects are listed by state and broken down into eight building types, including commercial, educational, housekeeping, and nonhousekeeping residential, medical, and government.   More than 5,000 projects are reported annually with an aggregate construction value of some $25 billion. About 60 percent of the projects are carried prior to the A-E selection or announcement of selection.

A companion service, the *Industrial Project Planned List,* carries projects in industrial, manufacturing, warehousing, and engineering categories. This list also costs $540 a year; both lists for $900 a year. The *Million Dollar Project Bid List,* from the same publisher, is a monthly listing of projects in the bidding stage and is therefore of primary interest to suppliers and contractors. The charge for all three Live Leads listing services combined is $1,200 a year.

BUSINESS DEVELOPMENT
REFERENCE SHELF

National Trade & Professional Organizations of the U.S.: Published annually by Columbia Books, Inc., Suite 601, 734 15th Street, N.W., Washington, D.C. 20005. Price $15.

Contains data on more than 4,700 national trade associations and professional and learned national societies. A key word index facilitates locating all groups under one industry or subject.

American Architects Directory: Published irregularly by R. R. Bowker Co., 1180 Avenue of the Americas, New York, N.Y. 10036, under the sponsorship of the American Institute of Architects. Price $37.50. Contains individual biographies and firm listings, with a geographical index. Helpful in locating potential associates or joint venture partners.

Dun & Bradstreet Reference Book of Corporate Managements: Published annually by Dun & Bradstreet, Inc., 99 Church Street, New York, N.Y. 10007. Contains biographical data on executives in 2,500 companies of major business and investor interest. Some 30,000 officers and directors are shown by company affiliation. Price on request to D&B.

Standard & Poor's Register of Corporations, Directors and Executives: Published annually by Standard & Poor's Corporation, 345 Hudson Street, New York, N.Y. 10014. Price $120.

Some 34,000 U.S. corporations are listed, with the names of officers and directors, annual sales, numbers of employees, and products. Standard Industrial Classification indexed. Brief biographies of 70,000 executives are in the back of the *Register*. Supplements are furnished subscribers in April, July, and October.

Congressional Directory: Published for each session of Congress by the Superintendent of Documents, U.S. Government Printing Office, Washington, D.C. 20402. Paperback edition $3; regular (hardbound) edition $5.50; thumb indexed edition $7.

The *Directory* contains an amazing amount of information about members of the current Congress, government agency executives, and the staffs of both. Indispensable.

U.S. Government Organization Manual: Published and revised annually by the Office of the Federal Register, National Archives and Records Service, GSA. Available from the Superintendent of Documents, U.S. Government Printing Office, Washington, D.C. 20402. Price $3 (comes only in paperback).

This book describes the functions, responsibilities, and staffs of the legislative, judicial, and executive branches. Also contains brief descriptions of quasi-official agencies and selected international agencies and lists many government publications. A companion piece to the *Congressional Directory*.

Who's Who in America: Published biennially by Marquis Who's Who, Inc., 200 East Ohio Street, Chicago, Illinois 60611. Price of the 37th edition $69.50.

Carries more than 80,000 biographies of important persons, including business and government executives.

Who's Who in the East: Published biennially by Marquis Who's Who, Inc. Price of the 14th edition $44.50.

Contains biographies of more than 20,000 persons in Connecticut, Delaware, Maine, Maryland, Massachusetts, New Hampshire, New Jersey, New York, Pennsylvania, Rhode Island, Vermont, West Virginia, and the eastern provinces of Canada. There are other Who's Who regional editions for the Midwest, South and Southwest, and the West. Also note *Who's Who in Finance and Industry* (18th edition $44.50) from the same publisher.

Polk's World Bank Directory—North American and International Sections: The North American Section is published each March and September; the International Section comes out annually in July. From R. L. Polk & Co., 2001 Elm Hill Pike, P.O. Box 1340, Nashville, Tennessee 37202. Both sections, $70 a year on a five-year contract. A single issue of each, $52.50.

Information about officers, balance sheets, and correspondent banks for more than 14,000 banks and almost 25,000 branches.

Dun & Bradstreet Reference Book: Published every sixty days by Dun & Bradstreet, Inc., 99 Church Street, New York, N.Y. 10007.

The *Reference Book* contains nearly 3 million listings for commercial enterprises in the United States and Canada. Listings are alphabetical by business name and by town within the state or province. Included in the listing is the product line and function, with keys to identify age of the business, its estimated financial strength, and corporate credit appraisal. The *Reference Book* is also available in sectional editions. Services include D&B's business information reports, provided under an annual subscription agreement. Cost of the service is determined by sectional area coverage and the number of business information reports included in the agreement.

Congressional Staff Directory: Published for each session of Congress by Charles B. Brownson, 111 So. Washington St., Alexandria, Va., 22134. Price $15.

Included in the *Directory* are biographies of almost 2,000 key congressional staff members, along with assignments of congressmen and their staffs to committees and subcommittees, locations of District and state offices of members of Congress, names of key personnel of Executive departments and agencies, together with liaison and information officers. An "Advance Locator" is published by Brownson as each new Congress organizes. Price of the paperback "Locator" is $3.50.

Contracting with the Federal Government—A Primer for Architects and Engineers: Published in 1969 by the Committee on Federal Procurement of Architect-Engineer Services, in cooperation with Federal Publications. Available from "Primer," Room 713, 1155 15th Street, N.W., Washington, D.C. 20005. Price $6.

Good basic book, although subject to outdating, on seeking and obtaining Federal A-E work, fees, and contract negotiation. A supplemental updating pamphlet is available free from the American Institute of Architects.

Who's Who in American Politics: Published biennially by R. R. Bowker Co., 1180 Avenue of the Americas, New York, N.Y. 10036. The 1973–74 edition is priced at $40.

Here is found biographical data on 15,800 political leaders and

government officials—city, county, state, and national. Indispensable to a Washington, D.C., office, but helpful no matter where a
firm is located.

MISCELLANEOUS SOURCES

Neither the listing of sources of specific job information nor the
recommended reference book section is intended to be all-inclusive.
Outside of a wide range of personal contacts, the list of which
should grow longer by each year's experience in client acquisition
work, these are the sources that successful firms have found to be
of most help in finding job leads and in assembling the initial
intelligence reports on potential clients. Some firms depend
heavily on *Dodge Reports* for job leads; others like *Thomas' Register* as yet another source of corporate information. The current
edition of the *World Almanac* is always an excellent general reference book to have at hand.

Some information sources seemingly fall between those giving
specific job leads and the publications with general and specific
reference material. One such difficult-to-classify source is *From
the State Capitals*, a series of forty different reports in newsletter
form covering everything from "Agriculture and Food Products"
to "Workmen's Compensation." Established in 1930, *From the
State Capitals* reports are compiled and published by Bethune
Jones, 321 Sunset Avenue, Asbury Park, New Jersey 07712. Of
primary interest to the design profession would be the weekly or
monthly reports on

Airport Construction and Financing
Highway Financing and Constructing
Housing and Redevelopment
Industrial Development
Institutional Building
Parks and Recreation Trends
Public Health
School Construction
Sewage and Waste Disposal
Urban Transit and Bus Transportation

These reports are issued monthly and priced at $36 a year, with the exception of Highway Financing and Construction, which is weekly and costs $72 a year.

Information about project newsletters and reference books covering foreign job markets will be found in Chapter 12, "Overseas Client Acquisition."

REFERENCES

[1] Royal Barry Wills, with collaboration of Leon Keach, *This Business of Architecture*, Reinhold Publishing Corporation, New York, 1941.

CHAPTER 5

Promotional Tools and Strategy: I
Brochures, Reprints, Books,
Motion Pictures, and TV

Fads and fashions exist in the promotion business, just as in women's clothes, architecture, and automobiles. For an example not related to selling professional services, some readers may remember when blotters were the staple items in promotions based on the constant reminder premise. When was the last time anyone saw an advertising piece printed on blotting paper? The reason for the disappearance of blotters (and pen wipers—an even older promotion institution) is obvious: the advent of the ballpoint pen in the 1940s.

All of which calls to mind the experience of the Ethyl Corporation, one of America's larger companies and the nation's major producer of lead compounds for gasoline. In one of its several efforts to diversify over the years, Ethyl turned to the manufacturer of fountain pens in the 1940s—just ahead of the introduction of the ubiquitous ballpoint. This marketing fiasco followed on the heels of Ethyl's decision to go into paper bag production, only to see the advent of the plastic bag industry.

Then, as the environmentalists gained stature and power in the 1960s, Ethyl came under heavy attack for its role in air pollution.

At about that same time government and private ecologists decided phosphates in washing detergents were a major source of water pollution. A chemical known as NTA (nitritotriacetic acid) was being touted as the ecologically acceptable substitute for phosphorus, so Ethyl and several other companies turned to the production of NTA. Government scientists then raised the twin specters of cancer and genetic damage as possibly being caused by NTA. Phosphates were back in, as less dangerous, and NTA was out.

If there is any moral here, other than the (fountain) pen is not always mightier than the sword of competition or that cleanliness is not necessarily next to ecologic godliness, it is that big fellows also make mistakes in marketing and promotion. The story of Ford's Edsel is too well known to need repeating here. The purpose of this chapter, like most of the remainder of the book, is to help professionals learn from the mistakes as well as the successes of others in the same field.

BROCHURES

Brochures may be either the most productive or the least effective items in the professional's kit of promotion tools. After reviewing several hundred brochures accumulated from architects, engineers, planners, and consultants over the years, plus a general familiarity with many others from the same groups of design professionals, one might venture these general observations:

1. Design professionals' brochures, by and large, are poorly done. Perhaps 10 percent are truly effective.
2. Many brochures come through as a kind of ego trip for the principals of the firm — and little more. In addition to the obvious narcissism, other frequently noted faults are wordiness leading to excessive length (and reader exhaustion), professional jargon, rambling and often meaningless expressions of design philosophy, and insufficient space (and thought) given to client interest about personnel and economics.
3. One of the secrets of good brochure production is to spend whatever is necessary to make an effective marketing tool for a firm and its services, without giving the client-reader the feeling that it is gold-plated. Conspicuous extravagance usually backfires, particularly where the client's project is modest and/or on a tight budget. It is also possible to spend a great deal of time and money on a brochure and have it turn out resembling the cheapest of discount store throwaways — but that is another matter.

Qualities to strive for in a firm brochure are relatively simple and should be evident to any professional: clarity, brevity, excellent graphics, plain straightforward presentation and text, and a demonstrated awareness of client interests.

Probably the most-used approach to getting a new brochure produced is to ask an advertising or public relations firm to put it together. The writer or writers and other specialists who will be assigned by the consultant to the brochure project usually will come to the assignment fresh, open-minded, and completely objective—in that this is the first time they have worked on a brochure for an architect or engineer. This is not said as criticism—design professionals take on unique or unfamiliar building types regularly —but as a warning that an office should be prepared to spend a fair amount of time over a period of one to six months with these specialists.

The writing of the text is the most critical aspect of putting a brochure together. Layout, graphics, picture selection, choosing type faces, and picking a suitable paper stock are all elements with which advertising or public relations consultants work every day, and these activities should require a minimum of the design professional's time and supervision. The consultants usually will not be familiar enough with the design profession to write well about a firm and its professional approach without heavy input and guidance from principals and others in the firm.

AVOID JARGON

Not long ago I sat in on a meeting in the conference room of a major architect-engineer-planner, where the third draft of the text for a new family of brochures was being reviewed. In 277 words of introductory material the word "problem" appeared eleven times. In each case "problem" referred in some way to the client and his project and the comment was made that few, if any, clients consider themselves a problem when they interview architects and engineers—yet this was the copy for a brochure whose prime purpose was to attract new clients. In the same introductory section the term "tradeoff analysis" was used. The writer had come out of the aerospace industry and reverted to systems jargon in this case. As the draft was discussed by the head of the firm, his design chief, the public relations man, and representatives of the marketing department, it became apparent rather quickly where the writer had gone astray in his approach and he left with enough input to do another rewrite.

The only real alternative to this system of "write and dissect," which is admittedly time-consuming, is for one or more members of the firm to write the initial draft and turn it over to a professional writer for polishing. A certain amount of conference time will still be required to arrive at a product generally satisfactory to all, but it can be minimized by proper planning and coordination.

It has been said that a brochure can serve as a calling card, a sales piece or a catalog—but never as all three. Brochures usually function as personalized calling cards, serving either as an attention getter or a reminder, depending on whether the client receives it before or following a visit. A brochure is not a substitute for a personal contact and, to my knowledge, no firm has ever gotten a job just by sending a brochure to the prospective client.

CLIENT USE OF BROCHURES

Four major corporate clients were asked how the architect's brochure figured in their selection of him for a job. The results of this minisurvey may be of interest.

Firm A agreed the brochure was a factor in its selection, adding that "personal contact, of course, was very important. . . ." This respondent also pointed out that he is building a library of brochures, illustrations, and articles about architectural and engineering firms who may be of future interest to his company.

Firm B's spokesman replied: "The brochure was interesting and informative, but it had nothing to do with our decision. We received it after we had already decided on the firm. We made our decision after inspecting his previous work and on the advice of our engineers."

Firm C: "Our familiarity with the architect predated [receipt of] the brochure. I don't really remember seeing it. But we had seen the firm's work and that sold us."

Firm D: "We don't pick architects by brochures, but the explanation of the firm's unique operation in its brochure was a big selling point. Our selection process was a long one. We first selected the twelve architects with the most prize-winning structures. Then we looked at their work. The whole process took three months and then the personal visits to the architectural firms began."

A subsidiary of firm D, incidentally, began its selection process by contacting the heads of its several divisions, which cover the entire United States. Each division head was asked to submit up to three names of design firms with whom he had had good results. Beginning with these twenty-one names the home office selection team narrowed the list down to three finalists, from whom a choice was made following interviews and visits to completed projects of all three design firms.

Another recent survey asked potential clients — municipal officers, developers, and realtors — what they looked for in planning firms' brochures. The interviewees were in agreement that they had never chosen a planner on the basis of his brochure — and never would. They are interested in the qualifications and experience of the firm's personnel: they want evidence that the firm is grounded in reality and is able to coordinate all aspects of pre-development planning; and they want to see completed projects of varying degrees of complexity and cost. All of the potential clients emphasized the point that they prefer the planner to document his scope of experience with a few well-chosen and related projects, rather than, as one client stated, "an indiscriminate mass of material thrust at the prospect in the hope that some of it will be interesting."

Observant readers may have noted the omission of any mention so far of an important specialized brochure, the Architect-Engineer Questionnaire, Standard Form 251, required by most federal agencies and by some state governments. The preparation and use of this brochure will be covered fully in Chapter 10. An official of the General Services Administration recently made the comment that while his staff is interested in a firm's Form 251, the selection committees with which GSA works are primarily interested in the regular A-E brochure, so both must be done well. He also had some suggestions about the organization of a firm's regular brochure. "Most A-E brochures," he observed, "run to heavy blocks of text in the front, with project photographs and descriptions relegated to the back of the booklet."

Put a *short* foreword or introduction at the beginning, then display the project pictures, he recommends. These should show completed jobs, he cautions—never model photographs and prefer-ably no sketches or renderings. Then add as many pages of text, biographies, philosophy, etc., as deemed necessary at the rear of the brochure. If a GSA staff man or a member of a selection com-mittee wants to take the time to read through the printed matter, he can—but he will already have been exposed to the important photographic record of a designer's practice in the beginning of the brochure.

BROCHURE SYSTEMS

A current trend in brochure design is to develop a system of smaller booklets, each devoted to one of the increasing number of dis-ciplines in medium- and large-sized offices. A general or "um-brella" brochure usually accompanies the specialized ones. Cover-ing folders or envelopes unify the set of loose brochures and are a part of the total brochure system.

The number of individual booklets in a system is limited only by one's imagination and budget. For example, in addition to the umbrella booklet, which normally will include architecture, a firm might have brochures covering planning, engineering, interior design, construction management, and computer services. A further refinement would get into separate brochures for selected building and project types. Airports, hospitals, university struc-tures, and high-rise office buildings are examples of this type of booklet. The aim in all of these approaches is to have as flexible a system as possible with preprinted, preassembled elements, sup-plemented by a minimum of custom-assembled materials.

NONSTANDARD FORMATS

While unusual shapes and sizes in format make their appearance on the brochures scene from time to time, the more-or-less standard 8 X 10 inch format is still the most accepted one. An odd-size brochure may require a large amount of waste of standard-size printing stock, resulting in higher costs. Press runs may be more costly because of extra time needed to position the plates. Most photographers are geared to producing 8 X 10 or 8½ X 11 enlarge-ments. Another argument for staying with standard sizes is that it makes the task of combining materials from several firms into a common brochure for associations and joint ventures much easier.

None of the foregoing is to discourage creativity or the use of nonstandard sizes and formats. But be prepared for higher production costs and complications in marrying your materials with those from other firms.

In designing the outside cover or an envelope of heavy stock to hold two or more brochures of a system, it is a good idea to place the firm's name on the folder so it can serve as a convenient identification tab in file drawers. If your covering folder is a least ¼-inch thick, the firm's name can be imprinted on the spine where it easily reads from above.

On the subject of nonstandard formats, a designer once explained that he had selected a jumbo format for a special study purposely to make it too big to file or conveniently throw away in a standard wastebasket. His rationale was that the client would take the easy way out and leave the report on his desk or a table—to serve as a continuing promotion piece for the consulting firm. It worked in this case.

Another type of brochure can be a helpful low-key sales tool during visits of potential clients to a firm's office. These are basically scrapbooks of clippings about the designer's work. Edward Durell Stone uses oversized red leather notebooks for this purpose, with his name printed in gold on the covers. Each book is devoted to clippings about one job—selected to be representative of a number of related projects from his office. Stone usually has three or four of these books at hand during an in-house presentation so he can pick one up and riffle through the pages at an appropriate moment or casually push the books across the conference table to the client for him to look through. These books never leave the office; they are too expensive to give to clients and, in most cases, would be difficult or impossible to duplicate. A shelf of twenty or so of the rich red leather-covered notebooks is rather impressive to many clients.

DEALING WITH PRINTERS

While many firms end up engaging outside professional counsel to put together their brochure, some offices are staffed and equipped to design promotion materials in-house. In these cases the staff member in charge of producing the brochure will deal directly with the printer. In selecting and working with printers, there are several points to remember.

Like architectural firms, printers come in many different sizes. There are neighborhood job shops specializing in small forms, announcements, stationery, and calling cards. There are commercial printers with ten or fewer employees who can handle some color work. And there are medium-to-giant-sized lithographers who do all kinds of general commercial printing.

Another way to bracket printers is by reproduction process. There are silk-screen specialists, short-run duplicating shops, letterpress printers, sheet-fed offset lithographers, web offset houses, and gravure printers. Some have combination facilities.

This might sound confusing, but a few inquiries will turn up printers equipped perfectly to handle any size or type of job. Within given categories of printers there are wide differences in quality. For the finest quality precision color work available, for example, one will know where to go after seeing samples of work done for others. Not surprisingly, superior quality is more expensive—and more effective—than average quality.

Printing estimating today is practically a science. In the better printing houses every operation is governed by job standards. Precise costs for all details are known and without all information and specifications for a particular job, the printer cannot arrive at an accurate cost estimate.

The Chillicothe Paper Company, in its publication *The Printer and the Buyer*, gives these ground rules to follow before calling in a printer:

- Broad Objectives: What must the job accomplish? Who will receive it? Answers to those two questions will help determine overall tone and design, which affect the cost. Another factor determining the cost is the delivery date. If a client waits until the last minute to start his job, overtime costs in its production may result.
- Physical Details: Size and number of pages, number of colors, quantity, type and grade of paper stock, self-cover or separate, folds, inserts, die-cuts, binding and envelopes.
- Contents: Number of photos and pieces of art (size and kind), amount of copy, kind of composition (display, text, machine, hand-set, etc.).

A few other tips and helpful hints for dealing with printers, based on an article in the *Public Relations Journal* for February 1972:[1]

- Printing the inside pages for some 8,000 copies of a two-color, 16-page brochure usually requires about six hours of press time. In Chicago and other large cities about $300 would be the charge for this service during normal working hours. If the job has to be run overtime on a Saturday, the client could wind up paying double that amount for the same work.

- Using four-color process photographs on pages 1, 8, 10 and 14 of a 16-page brochure with a press run of 10,000 copies could cost about $320 more than using the same color pictures on pages 1, 8, 9 and 13. The point here—if you are going to use full color in a brochure, carefully plan where it goes. A 16-page brochure typically will be printed on a 25- by 38-inch sheet of paper—one pass through the press for the eight pages on one side of the paper, another for the reverse side. For economy, plan colors to go on those pages that are printed at the same time.

- The economies of small run versus large run applies to the use of embosses, die-cuts, gate folds, end sheets and other such design elements. Like full color, they can add to the brochure's appearance; and they can also add substantially to per copy costs when the number to be printed is only a few thousand. At the very least, if it is planned to use such special effects in a brochure, find out what they will cost. The decision may be that it is not worth the extra costs.

- Instead of using justified lines (margins justified or lined up on both left and right sides of the page), consider having type set with uneven ragged-right margins. Minor changes often can be made in ragged type blocks during the last stages of preparation with a knife and paste, involving no expensive type resetting at all.

- The brochure designer should be at the printers when the pages come off the press, because adjustments at that time can sometimes eliminate the need for a complete rerun where the printing is unsatisfactory.

Once the ground rules have been established, make a layout or dummy to be used by the printer for his estimate. The more comprehensive the layout, the more accurate his estimate will be.

REPRINTS AS PROMOTION TOOLS

Many firms now include reprints of pertinent articles from magazines and newspapers with their printed brochures. The 8 X 10 inch format for the covering folder allows this to be done with ease; the nonstandard formats require some ingenuity to cope with the different sizes of enclosures. Whether or not reprints are inserted in the brochure package, they are always a good general sales tool for the professional. Something in the human psyche tells one that if it has appeared in print somewhere it has to be true.

Since most articles about a firm's work will be carried in the architectural trades ("shelter" publications), a potential client seldom sees the piece in its original publication. Other stories about jobs may appear in specialized—even obscure—publications seen by only a few thousand readers with special interests. Some examples of these small or controlled circulation, special-interest magazines are those published by and for associations of bankers, school administrators, auditorium managers, ministers, and police officials.

Any and all published articles should be considered as potential grist for the total promotion kit. Reprints can usually be obtained from the publisher if he is given sufficient advance notice of requirements. Or one can arrange to have almost any article reproduced by a local offset printer. When color is involved in the original article reprints should be ordered directly from the publisher. He will keep the forms for the article intact and print the reprint order following his regular press run. A special cover—called a "self cover" in the trade—can usually be included in the instructions to the printer, with the name and date of the publication and the design firm's name on it.

Prices for reprints will vary so there is no point in trying to give any guidelines here. As general guidance, 1,000 copies cost little more than 250 or 500 copies, particularly when it is a full-color, multipage reprint. Don't worry about what to do with 1,000 copies of an article; if only 100 copies are used in two years' time and two or three of them are instrumental in getting an interview, the investment was worthwhile. And a use may well present itself for all 1,000—say at a major convention where the firm has a model on display. Remember, once the forms come off the press it may be impossible, and it certainly will be impractical from a cost standpoint, to reorder.

Black-and-white copies, as was mentioned earlier, do not pose such a problem. As long as one good clean copy of the article is retained, new offset plates can be made and any number of additional copies produced quite reasonably. Naturally, permission to reprint should always be requested from the original publisher.

FOREIGN PUBLICATIONS

For some reason, few American architectural, engineering, and planning firms take advantage of the interest foreign publications —architectural and others—have in U.S. firms and their work. Editors of Italian, German, French, and Spanish architectural magazines use such material and in some cases pay for it.   One of the Spanish architectural magazines, following publication of a story about a corporate headquarters building in New Jersey, notified the architect that the publisher had deposited a thousand or so pesetas to his account in a Madrid bank, where it could be claimed in person.   Since it was the equivalent of about $16 he didn't bother to make a special trip to collect—but it was a nice gesture.

Unfamiliarity with foreign language need not be a concern, incidentally, since the editors take care of the translations.   Correspondence will usually be in English as well.   See Figure 5-1.

A foreign language reprint or two, included along with English articles, has been known to make an impression on certain clients.

BOOKS BY PROFESSIONALS

Still on the subject of writing, very few architects, engineers, and consultants have taken advantage of one of the best possible promotional tools, authorship of articles and books.   Many more professionals write for newspapers and magazines than attempt to write books, of course, but the total number of creative writers in the profession is still pitifully small.   Our old friend from earlier chapters, Royal Barry Wills, wrote eight books in the sixteen years between 1941 and 1957, not counting the work of revision on two of them.   Edward Durell Stone has produced two books on his work and a few other architects have somehow found the time to turn out one or more books.

At one time "vanity publishing" was a dirty word.   Today, some design firms feel they can justify the costs of publishing their own books in that it gives them an excellent public relations and marketing tool.   If a firm is large enough to place a prepublication order for at least 5,000 copies of a book, even the old-line commercial publishers are interested, everything else being equal.   As a rough guide for this type of book, 5,000 copies is often the break-

Fig. 5.1 *Representative articles from foreign language publications.*

even point for a publisher. Given a reasonably well-written text, good illustrations, and the guaranteed 5,000-copy order, many publishers will be interested in discussing the matter.

DOUBLE-DUTY REPRINTS

"Piggybacking," to some public relations practitioners, is the practice of squeezing every possible promotional drop out of any one article or situation. For example, one of a firm's principals might address a group on some fairly newsworthy subject (first publication). If properly merchandised this talk will be covered in the local or even regional press (second publication). The entire text of the talk might be picked up by a magazine (third publication). Then, a friendly congressman could read the magazine article into the *Congressional Record* (fourth publication). While this doesn't begin to exhaust all of the piggybacking possibilities in this case, it is evident that the original material has had three good additional exposures beyond the speaker's original audience. The total audience thus reached could be in the tens or even hundreds of thousands, rather than the original 100 or 200 people who may have heard it in person.

A few more words on the *Congressional Record* as a promotion tool for professionals. Regular readers of the *Record* will know that many of the back pages of each issue are usually taken up by a potpourri of material which various congressmen have been granted permission to "read" into the *Record*. Obviously, they don't really read this mass of material on the·floor of the House and Senate. It is only necessary to have one of their aides deliver it to the House or Senate clerk, who adds it to the actual record of the day's deliberations and it ends up printed as part of the day's official record.

Getting a quantity of reprints of the appropriate pages of the *Congressional Record* for distribution to clients and potential clients is highly recommended. The best types of articles to suggest to your congressman are those from architectural and engineering magazines giving a general overview of your firm and its operation or a story about a particularly significant design solution on a major project.

Needless to say, the *Record* is not open to design professionals on a will-call basis. No real rule of thumb can really be given,

Congressional Record

United States of America

PROCEEDINGS AND DEBATES OF THE 91ˢᵗ CONGRESS, SECOND SESSION

Vol. 116   WASHINGTON, THURSDAY, FEBRUARY 5, 1970   No. 15

Senate

The Senate met at 12 o'clock meridian and was called to order by the President pro tempore (Mr. RUSSELL).

The PRESIDENT pro tempore. It is the pleasure of the Chair to present to the Senate as guest chaplain today the Reverend James P. Wesberry, D.D., pastor of the Morningside Baptist Church in Atlanta, Ga.

The Reverend James P. Wesberry, D.D., offered the following prayer:

Gladden our lives, O God, our Father, with the light of Thy redemptive purpose. Cleanse us, we pray, from all evil. Open our hearts to Thy love which satisfies our deepest need and to Thy strength which matches our heaviest burdens. Grant that we may move in the performance of our duties as the unhurried stars in the orbit of eternity, without haste or confusion, but always with shining steadfastness. When faced with obstacles bigger than we can handle, may we find within us a spiritual power that breaks through, and when worldwide responsibilities mount upon us, may we go forward with the sureness of the mighty river that runs its destined channel to the sea.

Through Jesus Christ our Lord. Amen.

MESSAGES FROM THE PRESIDENT

Messages in writing from the President of the United States submitting nominations were communicated to the Senate by Mr. Geisler, one of his secretaries.

EXECUTIVE MESSAGES REFERRED

As in executive session, the President pro tempore laid before the Senate messages from the President of the United States submitting sundry nominations, which were referred to the Committee on Armed Services.

(For nominations received today, see the end of Senate proceedings.)

THE JOURNAL

Mr. MANSFIELD. Mr. President, I ask unanimous consent that the reading of the Journal of the proceedings of Wednesday, February 4, 1970, be dispensed with.

The PRESIDENT pro tempore. Without objection, it is so ordered.

LIMITATION ON STATEMENTS DURING TRANSACTION OF MORNING BUSINESS

Mr. MANSFIELD. Mr. President, I ask unanimous consent that statements in relation to the transaction of morning business be limited to 3 minutes.

The PRESIDENT pro tempore. Without objection, it is so ordered.

HON. HUGH SCOTT
OF PENNSYLVANIA

IN THE SENATE OF THE UNITED STATES

Thursday, February 5, 1970

Mr. SCOTT. Mr. President, I invite attention to a Philadelphia architect who has designed Philadelphia institutions and whose work has been recognized all over the Nation. While he has designed projects outside his home city, Vincent Kling recently reached the pinnacle when he was selected to design the new Federal Triangle here in the Nation's Capital. Building Construction magazine's latest issue contains a complete profile of Vincent G. Kling. I ask unanimous consent that the article be printed in the RECORD.

There being no objection, the article was ordered to be printed in the RECORD, as follows:

VINCENT G. KLING AND ASSOCIATES—THE OFFICE OF AN ARCHITECT WHO FOCUSES 500 ARCHITECTS, PLANNERS, AND ENGINEERS ON DISTINCTIVE BUILDING DESIGN AND THE TOTAL OWNER PRODUCT

The practice of architecture always seemed to require, classically, and at its best, the dominance of one creative man. Today, an architectural firm, motivated still by that ideal while handling substantial and complex contemporary architecture, is Vincent G. Kling and (his) Associates. Their distinctive work is the result of a man blessed with sufficient drive to guide, almost personally, the work of a staff grown to nearly 300 in its 23rd year.

If you ask Vincent Kling to characterize his firm, he starts at the beginning with orderliness:

"When we build, we are telling the world what we stand for. Our structures will influence our lives for a very long time. The choices we have with which to achieve this influence are legion; no longer is it a simple matter of bricks and mortar, windows and doors. The new methods, new systems, and a seemingly endless demand from more and more people give the designer fascinating opportunities at every turn, as we enter a building surge which, in 30 years, will witness the certain doubling of our shelters."

"The architect, a generalist by training and practice, plays the major role in the conception and execution of the design of spaces, places, and enclosures. He leads and directs a wide spectrum of specialists.

Noting the particular demands of today, Kling states: "Our fundamental conviction in approaching the design of every project is that architecture is for people, not just architects. With increasing urban concentration and megalopolitan sprawl, the greatest challenge to the architect is to recreate environments for people, within and around his structures. Elegance, grace, style, functional efficiency, economy and durability are still as important to owners as ever, but today the most pressing need is for humane spaces in which people can live and breathe.

"This, of course, makes our task more complex. Our office offers a comprehensive service from research, programming and planning, land utilization, and movement systems to finished engineering design, construction and final inspection. This includes landscape and site preparation, interior design, space planning, communications, cost analysis and budgeting."

A more classical recital of the role of the architect for these years could hardly be composed. It would serve of course for any sophisticated full-service team of men, but for the Kling office, in particular, it describes really the thrust of Vincent Kling himself. Here is one of the few architectural sole proprietors of today with such a tremendous talent for personal organization and with such command of his staff, that he is able to reach deeply into the critical decisionmaking on any project, and earn the right to point to most of his buildings and say, "I was the architect." To be sure, without a certain pattern of capable, understanding, and supporting associates, he would be powerless. Yet, to them, he remains their ultimate source of unique directional power.

Kling can point to a surprising number of buildings right in the front yard of his office which is in the heart of Philadelphia (although his work spreads over the states of eastern U.S.): the Municipal Services building across from the venerable old City Hall, the realty of Penn Center, the IBM building, John F. Kennedy Plaza, and eight other buildings or courts. Upcoming are the twin towers of Center Square, and not far away is the new U.S. Mint. In a city noted for its pride in Independence Hall, its traditions, and its cultural attainments. This manifestation of confidence is enviable indeed.

BEAUTY STILL SUPREME

At the heart of Kling's strength is an intense dedication to design. In these architectural years, when exploding technology in structure, materials and methods has almost stolen the prime attention of architects and engineers, when client demands as well as labor has upped the cost of buildings, Kling still insists that while these demands

---

Fig. 5.2 *Congressional Record with article on Vincent G. Kling and Partners.*

but insertions should probably not be requested more often than once a session—and even that is pushing it a bit. Naturally, a great deal depends on an individual's personal relations with his senators and representatives.

In reproducing material from the *Record* some firms have been known to do a paste-up of their article on the cover page of that day's *Record,* following the chaplain's prayer. This treatment tends to add importance to the material. See Figure 5-2.

OTHER OUTLETS FOR REPRINTS

When looking for the more exotic outlets for articles don't forget the United States Information Agency (USIA), our official voice in foreign countries. The USIA has several regular publications which go to all embassies and consulates for use of foreign-based information officers. These publications contain photographs and stories which can be translated and furnished to local magazine and newspaper editors. In some cases the embassy puts out its own magazine in the language of the country in which it is located. Significant buildings, new design and construction techniques, and award-winning structures of many types are possibilities for this treatment. Here, too, the primary consideration is to get copies of the publication so reprints can be made for further distribution. USIA's Press and Publication Office is in the Agency's headquarters building, 1776 Pennsylvania Avenue, N.W., Washington, D.C.

MOVIES THAT SELL

Many design firms have discovered the advantages of motion pictures in marketing. "The Architect as Film Maker," in the *AIA Journal* for February 1971, pages 23–25, is one of several good general articles on the subject. One nice thing about motion pictures as a promotion tool is that their production cost usually can be included in the client's budget. If copies would be useful in a firm's marketing efforts, they then pay only the charge for duplicating the finished film. In such cases it might cost as little as $150 per print for a fifteen-minute film which cost upwards of $20,000 to make.

When the client is a governmental body it is sometimes necessary to produce a film to sell the project to the voters in advance of a bond election. This point is covered thoroughly in "The Architect as Film Maker," using the Hartford, Connecticut, Civic Center as a case study.

Occasionally it is worthwhile to consider producing a special film to use in an important interview. These can be tricky and the full cost comes out of the designer's pocket. The *AIA Journal* article already cited covers the making of such a film. This one was for an interview to select an architect for the new student union building at Syracuse University. For a number of good and compelling reasons it was decided to build the film around interviews with the ultimate client-users—the students. Almost documentary in style, the body of the film consisted of semistructured interviews with students of a variety of views: personal, political, and philosophical. As usual, some of the film footage which ended up on the cutting room floor was the most interesting. One of the leading questions the architect-interviewer began with was, "What is the one thing about your life at Syracuse University which concerns you more than anything else?" One baby-faced coed, after identifying herself, looked straight into the camera and answered, "My main concern is that I'm pregnant and my mother doesn't know it." Another sweet-faced miss took up four minutes of camera time in profane descriptions of the college administration, her instructors, and her roommate, using four-, five-, and six-letter words which would have made the most hardened Marine blush.

These incidents are no problem, since the footage is easily excised out in editing. This story has a kind of O. Henry conclusion. The film was shown during the interview and since all of the interviews were open to the public and word of the film making on campus had spread rapidly, some 150 students were packed into the relatively small hall. At the end of the film the architect received an ovation of cheering and clapping lasting for several minutes—a first for interview presentations in his experience. The university development director begged for a copy of the film to use as his primary fund-raising tool for the new student union, a request which was granted on the spot. Two weeks later the architect was informed that the job went to another firm. That firm, incidentally, gave a standard presentation of talk and 35-mm slides.

Another possible film tool for architectural and engineering firms uses the filmograph process. This involves a simulation of movement from still photographs and slides and was quite popular as a television feature a few years ago. By devoting a second or less to each image, one can cover the Civil War, say, in two minutes. The process is subliminal in effect and is admittedly somewhat of a visual stunt. A design firm could profitably adapt this technique to show a fast run-through of their past work—

say 200 major jobs—as an opening to the interview, then switch to the more conventional slides or flip charts for the balance of the presentation. As in anything else, the filmograph opening would soon become trite if every firm used it, but I know of no office now employing it.

Before getting into motion picture production on any real scale, there are several important points for the design professional to consider. Some of these were enumerated in the *AIA Journal* article mentioned earlier:

Choose your film maker the same way you would choose any other consultant. Look at his work, talk to him, talk to some of his other clients. But above all make sure that you end up working with someone you are convinced has a feeling for architecture and what you are trying to accomplish. Make sure also that the person you are talking to is the one who will be doing the work. Don't hesitate to ask if he is the person who is going to be working on the project three months later. You need this kind of continuity. The job also demands availability, since a film tries to fall apart at every opportunity.

Usually it is more economical to bring in a professional film producer than it is to try to assemble your own film-making staff and equipment. Cameramen, soundmen, and other technicians are important, but the key people are the director and the producer. Both should be sympathetic to what you want to get across through the film, and both should be willing and able to devote all the time necessary to the project. So, by the way, should your own "producer"—your staff member responsible for the film. He shouldn't be afraid of screenings scheduled at odd hours, or of redoing everything at the last minute.

Another important factor is to have the right people in control of the film's production. The film maker has been engaged because he is a professional in his field; creative control should be left to him, with your approval and/or guidance.

About money: A good rule of thumb for a professionally made 16-mm color/sound film is $1,200 to $1,500 per running minute. Some run as high as $2,000 a minute. But don't let those figures scare you—that's for the more elaborate film, using the most expensive equipment. It can be done for less.

Much depends on the way the film is to be used. A presentation aimed at persuading corporate or government decision makers will call for a different tone and style than one aimed at influencing students. Techniques that will reach one group won't necessarily reach, and may even antagonize, the other. For a film that is to sell a project or an idea, you may want to spend the money for a thoroughly professional presentation; for another type of film, you can experiment with less costly methods and techniques. The important thing is not how much money you spend, but what you get for it.

Some sort of agreement should be reached with the client (if you are doing the film for a client) about the cost; in any case, a cost ceiling should be established.   There always seems to be extra costs cropping up, and it is a good idea to identify them early and figure them into the budget.   These extras can include such items as stock footage, snorkel work, animation, and location shooting.   If they aren't figured in from the start to give a realistic cost ceiling, the estimated maximum cost can be just the beginning.

And make sure everything is cleared.   Models, actors, music, quotes from existing works—all should be cleared ahead of time to prevent embarrassment and surprise charges later.   With a reputable film maker working with you, this should present no problems.   He will take care of it himself.[2]

Closed-circuit television presentations, live or utilizing video tape, are related to motion pictures as a promotion tool.   Video-tape production and playback equipment has dropped greatly in cost from a few years ago and some of the larger firms are using television routinely for in-house model studies, rehearsing principals in new business presentations, and for part of the actual presentation.   Immediacy is one of the advantages of closed-circuit TV; the small receiver screen size is a disadvantage of the medium.

REFERENCES

[1] Joseph W. Kutchin and Donald Levy, "How to Keep Your Annual Report Costs in Line," *Public Relations Journal,* New York, February 1972, pp. 19–22, 54.

[2] Vincent G. Kling, Jr., "The Architect as Film Maker," *AIA Journal,* Washington, D.C., February 1971, p. 25.

CHAPTER 6

Promotional Tools and Strategy: II
Correspondence, Proposals,
Job Histories, and Other Tools

One of the most important but least considered aspects (in many offices) of the business development process is the correspondence incidental to any job acquisition efforts. With few exceptions, a minimum of three letters should be written during the period following the first contact with the potential client and his awarding of the contract for the design work. In order, these are:

1. Initial expression of interest
2. Proposal letter—usually following the interview
3. Letter of thanks (or regrets) following the contract award

These are the minimum requirements; in many cases the total correspondence file for a prospect will contain many times this number.

There is no *best* style for writing any letter to a potential client —otherwise one firm theoretically would get all of the clients. It is usually assumed that a principal has an adequate enough grounding in the niceties of the English language to handle the composition of a letter in a manner to reflect credit on his firm and his associates. Unfortunately, this is not universally true

106

and clients tend to shy away from practitioners who cannot spell or who become enmeshed in convoluted syntax.

SAMPLE LETTERS

Because there is no ideal style for writing a letter of interest or a proposal, one hesitates to include "sample" letters in a book such as this for fear that they may be utilized in toto and with no regard for a specific client's project and requirements. However, to illustrate how *one* writer approached the problem, here are examples of a one-page and a two-page letter of interest:

Mr. Roger J. Thomas
Superintendent of Schools
933 Opal Street
Bratsford, Massachusetts 00000

Dear Mr. Thomas:
   This is in response to your invitation to submit my expression of interest in being considered as the architect for the proposed school building in Bratsford.
   As you may know, we are quite familiar with Bratsford and Perkins County, having served as architect for the Morton Company's Research Laboratory some three years ago. Currently we are involved in a major facility for National Electronics about 90 miles to the east of you—the informations systems complex in Maxwell. Still in Massachusetts, we are doing the new Academic Building at Amherst for the University of Massachusetts.
   Some other representative New England projects in which we are involved are the new Civic Center in Farmdale, the Health Center for the University of Maine's Medical School in South Depford, and new administration, production, engineering, and studio facilities for television station WORI-TV in Drayton, Rhode Island.
   During the past eighteen years my office has completed some 400 projects, totaling more than 25 million square feet of building space with a total value of almost $1 billion. Our school design work represents around 15 percent of this total.
   We offer an extremely broad spectrum of services related to the design and planning processes. In addition to our urban planning, architectural and engineering design, documentation, and construction management and supervision, our 250-man firm has extensive experience and capabilities in related areas such as large-scale land and site planning, programming and space analysis, research, interior design, landscape design, cost analysis and budgeting, and communications. All these extended services are available to our clients.
   I and my senior associates believe that your expressed desire for "a

creative school for total community use" would offer us an opportunity to focus our many years of directly and indirectly related experience on the development of a truly satisfactory and productive solution for Bratsford.

I look forward to the opportunity of exploring the subject in detail with your selection committee. In the meantime, I hope the enclosed brochures will give you some insight into our experience, abilities, scope of past and present work, and personal philosophy regarding the design process and architect-client relationships.

Very truly yours,
Arthur K. Smith, AIA

Dr. Elroy R. Marlborough, Vice President
University of Topeka
Topeka, Kansas

Dear Dr. Marlborough:

This is in reply to your invitation of April 00, 0000, to submit appropriate material supporting our interest in being selected as architects for the new central library building at the University of Topeka.

My initial expression of interest, in late October 19—, covered certain of the points to be considered by the University Building Committee in their selection process. I am enclosing a copy of that correspondence as part of my supporting portfolio.

First of all, I wholeheartedly agree with the University Building Committee's intention to proceed with the design for the complete library, with provisions to stage the construction in line with availability of funds. We have done this in a number of projects of major importance; currently our new headquarters facility for the International Assistance Program in Washington, D.C., is progressing on this basis. In this case, delay in land acquisition is the reason for staged development, but I know from experience that it more often results from funding restraints. I would anticipate no problem in arriving at optimum staging steps in the design and execution of the library.

On a related point, I am wondering if you are considering commissioning a comprehensive master plan of the campus concurrent with the design of the central library? If the new library is to be the focal point of future development of the total campus, it would be well to involve the planner at an early date. Since our planning division has accomplished a variety of master plans for universities and other clients, I have rather strong feelings on the timing of such plans to benefit all concerned.

In any major building project, such as your new central library, I believe that communications between the architect and client must be of the highest order throughout all stages of design and construction. For this reason, our usual approach is to assign a senior partner as principal architect, with one or more staff architects on the team, which you may be certain will include some of our outstanding University of

Topeka graduates. I involve myself deeply in the programming and concept stages along with these senior men, and maintain a close relationship with the team throughout preliminaries, schematics, construction documentation, and construction. The point here is that our clients deal with the same partner-in-charge and his technical team from beginning to end of a project.

Another point on the subject of architect-client communications concerns our association with an architectural firm in the Topeka area. If the University Building Committee is in general agreement that such an association is desirable I would plan to select a local firm after consultation with the Committee, so that a mutually satisfactory team may be established for your central library project.

Since I referred to the broad spectrum of services we offer in my earlier letter, I am enclosing in the portfolio a copy of our current organization chart. This will better identify for the Committee our six areas of primary competency: planning, architecture, interior design, engineering, computer sciences, and development services. Although these are all essentially independent divisions, under our centralized-management concept staff members are drawn from each division as the project warrants. In this manner we set up a total-service team tailored to the requirements of any given project.

I hope that the accompanying material, in addition to what you already have received from us, will convey to you and the members of the University Building Committee the wide variety of projects and clients we have been privileged to serve over the past eighteen years. While I tend to agree with the Committee, in their feeling expressed in your last letter, "that prior design of a university library is not the sole determination of architectural ability for this project," I am very proud of the university libraries we have done to date.

I am looking forward to the opportunity to present our credentials in person to the Building Committee. Anticipating your desire to become more informed about our performance capabilities, I am attaching a list of clients whom we have served on projects of related character and scope. Please feel free to contact any of them for a direct reference.

If there is any way in which we may be of service to you in the interim, please call on me.

Sincerely yours,
Arthur K. Smith, AIA
Enclosures

The architect made several important points in the second letter (to Dr. Elroy R. Marlborough). In the third and eighth paragraphs he engaged in feedback from Dr. Marlborough's letter of invitation. Always look for points in the letter of invitation which appear to be of particular interest or importance to the potential client, rephrase them, and feed them back to him in your reply. This will prove that you at least read his letter carefully. In the fourth paragraph the architect makes a fairly subtle pitch for an additional job—a campus master plan centering around the new library. It is possible that the importance of a master plan has not occurred to the university officials. Even if they decide against it the architect should score some points for thinking about it. In his ninth (next to last) paragraph the architect mentions his enclosure of a list of past clients. Most potential clients eventually will ask for such a list so why not volunteer it first? This technique also gives a professional the opportunity for continuing contacts with his past clients. Permission of the previous client is normally asked before giving him as a reference and this simple request has been known to bring in another job from the past client as well as leads on other work he has heard about.

PROPOSAL LETTERS

The straight proposal letter, as opposed to the letter of interest, can be a bit more involved since it gets into the area of contracts with the attendant legal implications. In the example below, a three-page proposal for a college dormitory, note that the architect has already assigned the prospect a job number. This could be considered a marketing gimmick, but it can be effective. Also, this proposal is designed to serve as a contract between architect and client. Some of these letters may actually be signed by the client as a binder on the architect, but this is rare. The negotiation of a design contract can be a long and wondrous process and few attorneys would allow a client to enter into an agreement on the basis of a proposal letter. Then why use this format? It is another form of marketing strategy, to demonstrate to the potential client that you are dead serious about becoming his architect; you are ready to proceed *now*.

Mr. Carson F. Clatworthy
Vice President—Development
Rustin College   Reference: Rustin College
South Falls, North Carolina   Job No. 000

Dear Mr. Clatworthy:
    We are most pleased that you are considering us as your architects for the proposed dormitory complex at Rustin College. We have enjoyed our past associations during the design and construction of the chapel, science center and the master planning study, and we are most desirous of continuing to serve Rustin.

You mentioned that you would like us to submit a proposal wherein we would perform a scope of services for studies up to and including completed preliminaries on a man-hour basis. Our proposal for these services follows.

We would assist you or your designated representative in firming up a written program of space requirements. Using the program developed for the proposed buildings in the master plan as a guide we would revise it as required to arrive at a scope of work.

Based on the approved program, we would prepare preliminary studies incorporating into the building necessary structural, mechanical, and electrical systems and other specialties such as food service, in order to properly scope the project for preliminary estimates. We will prepare in conjunction with the studies a cost estimate and outline specifications. The preliminaries would define the building as to number of floors, building mass, relationship to existing buildings, and exterior expression. It will be necessary to take as built measurements of existing buildings, as they may affect the outcome of the design.

For the services mentioned above we would request reimbursement based on billable technical payroll times a factor of 0.00. The compensation is developed as follows:

The billable technical payroll includes the cost of principals and employees engaged in the work, including architects, engineers, designers, architectural staff, draftsmen, and specification writers in consultation, research, designing, producing drawings, specifications, and other documents pertaining to the work.

As used herein the cost to the architect of the billable technical payroll will be determined under a man-hour formula in steps as follows:

1. The number of man-hours expended by architect's professional and technical personnel and principal consulting engineers in providing architect's services hereunder will be multiplied by the straight time rates of pay for such personnel, but such rates shall not exceed $00.00 per hour.

2. The results of step 1 will then be multiplied by a factor of 0.00. The man-hour formula provides compensation to the architect for all personnel expense such as salaries and employee benefits and for overhead and profit, but it does not include other reimbursable expense or fees for which separate provision is made.

3. Payment for overtime authorized by the owner will be determined as follows: The total number of man-hours expended in overtime will be multiplied by the straight time rate of pay for personnel thus engaged, the product obtained to be factored as above. The overtime premium payment will be determined by multiplying one-half of the total number of man-hours expended in overtime by the straight time rate of pay.

In addition to the direct labor costs incurred, there can be reimbursable expenses incurred by the architect and they would be as follows:

Reimbursable expense will be actual expense incurred by the architect properly incidental to his services hereunder including, but not limited to, the expense of the following:

1. Transportation and living costs incurred by the architect when traveling in discharge of his duties hereunder on other than trips in and about Chicago.

2. Any special consultants, engaged with the approval of the owner, such as kitchen and acoustical consultants.

3. Other disbursements on owner's account approved by the owner.

4. Special presentation materials such as colored renderings, detailed architectural display models of the project, photographs, and printing.

5. All telegrams and telephone toll calls required in connection with the services rendered hereunder.

6. Reproduction of drawings and specifications (excluding copies for the architect's office use and triplicate sets for each phase for the owner's review and retention).

7. Fees paid for securing approval of authorities having jurisdiction over the project.

Since the services performed are on a time-and-material basis the contract may be terminated at any time, reimbursing architectural services rendered to that point. For your cost control we will notify you when we expend $00,000, and request written approval prior to continuing design services. We would like to perform complete architectural and engineering services through to completion of construction for this most exciting project and would credit applicable portions of the fees received for services during the preliminary stage against an agreed upon lump sum or percentage fee.

If you concur with the services as set forth in this letter proposal we would appreciate your so indicating by signing your acceptance below.

Very truly yours,

ARTHUR K. SMITH & ASSOCIATES

Max Fortran, AIA, Associate
Accepted this _____ day of _____, 19—.
Rustin College
By: _____

One more example of correspondence with a potential client— this one a rather lengthy excerpt from a five-page, very detailed letter to a client who demanded an expedited design and build schedule. He wanted beneficial occupancy of his 100,000-square foot building within eighteen months. Computer facilities occupied about one-half of the total gross area. The writer of this letter coined a Washington-sounding acronym, "FTD/C," for what the client wanted—fast track design and construction:

Three examples of fast track design and construction (FTD/C) by our office should demonstrate our capabilities in this field:

1. The design and construction of the World Communications Corporation color television picture-tube plant in Newton, Ohio, necessitated a detailed master plan for an undeveloped and exceptionally difficult site. The plant had to meet very exacting requirements and be constructed with maximum speed for full production at the earliest possible date.

WCC needed a plant totaling 200,000 square feet, including warehouse space and cafeteria wings and separate administration area, for the manufacture of 500,000 color television tubes annually; a large percentage of world production of this item. Harold F. Jones & Associates' concentrated design and construction schedules succeeded in getting WCC into production of the first color picture tube in only eleven months.

In addition, the plant had to be designed as a general electronics facility, readily adaptable to different product requirements, since there are constant changes in electronic products and product engineering.

2. In 1960 the mushrooming market for molecular electronic devices prompted the Southern Falls Electric Corporation—a pioneer in the field—to build a new facility in Lowden County, Virginia, in which to consolidate and expand its East Coast research and production activities, then scattered in several locations.

Due to the pressures of new business, Southern Falls wanted the first phase of the new facility to be ready for initial occupancy in early 1962, a scant eight months after commissioning our firm to design the complex. To assure that the structure would keep pace with the burgeoning market—calculated at that time to have a multimillion dollar potential within five years—the facility had to be designed so that it could be enlarged easily and economically in relatively small increments to an ultimate size of ten times the initial area.

Due to the unique problems in producing molecular electronic devices—very small blocks of materials which microminiaturize functions performed by electronic circuits or even whole systems—all research and manufacturing areas had to have constant temperature and relative humidity. These areas were also air conditioned with "super clean" air free of all particles over .3 microns, so that the environment within all laboratory and manufacturing areas would have extremely low dust levels. Finally, the company required provisions for installing completely equipped clean rooms in selected areas in the future.

The first phase of the project, completed in early 1962, contained 100,000 square feet and provided space for 550 persons. To keep pace with the projected demand, the master plan provided for expansion up to 800,000 square feet to house 4,000 research, production, and administrative personnel.

That design considerations do not have to be sacrificed to achieve accelerated construction and early occupancy by the client is attested to by the many awards given the Southern Falls Electric Corporation facility. These honors include *Factory Magazine's* "Ten Top Plants of 1963" award, the American Institute of Steel Construction's "1965

Architectural Award of Excellence," and the national Honor Award of the American Institute of Architects.

3. The National Electric Corporation's Missile Technology Center at Moss Creek, Alabama, was occupied fifteen months from start of design.   This project demonstrates how an architect experienced in meeting client schedules, and with the necessary staff size and backup specialists in-house, can telescope design and construction of a multi-phase complex building program into a brief time span without sacrificing the original design concept.   The architect also must be able to accommodate the client's day-by-day revisions, which occur of necessity in projects of this type.

Phase I was 300,000 square feet; Phase II, 325,000 square feet.   The design achieved a high degree of flexibility, including future expandability.   Major functions were zoned separately for flexible subdivision and expansion.

The complete initial submission to a potential client usually will consist of a letter of transmittal, in which interest in his project is clearly expressed, as in the preceding examples; one or two printed booklets and a custom-assembled qualifications brochure highlighting the firm's experience in projects relevant to the one under consideration by the prospective client.   The qualifications brochure, in addition to the obvious materials, might include reprints of pertinent articles and copies of the company newsletter, if one is published, with articles on similar assignments.

The final piece of correspondence between design professional and the potential client, the letter thanking him for the interview, will be covered in Chapter 8.

JOB HISTORIES ARE IMPORTANT

Job histories may seem to be a mundane subject for our consideration at this point, but few in-house records are so universally neglected by architects, engineers, and planners, and job histories can play an important role in promotion and marketing.   One recurring problem is to develop a format which will serve the needs of all within a firm: administrators, planners, programmers, designers, draftsmen, accountants, specification writers, estimators, and the promotion and marketing staffs.

Architectural Record's Specification Outline, developed over several years and first used in 1971, is recommended as a guide for setting up a job history format in architectural and engineering firms.   It can be adapted to the use of planners and a wide range of consultants, as well.   The two inside pages of the cover folder, reproduced in Figure 6-1 on the next two pages, consolidate the most important information about a project.   A five-part series of separate booklets, contained within the cover, are for general specifications; structural system; heating, ventilating, air conditioning; plumbing; and electrical system.   Since each major division of the Specification Outline is complete within its own booklet, the persons responsible for the individual disciplines may fill out their sections simultaneously.   When all parts have been completed and reassembled in the cover folder the job has been thoroughly covered.

Accurate and complete job histories are important in compiling information required for AIA design competition entries and for many other design awards.   The public relations staff also requires much of the information in a comprehensive job history when writing releases about a project.   And new staff members can get a fast orientation to the firm's important work by reviewing job histories.   So, whatever format is decided upon, make certain that job histories are being compiled regularly, accurately, and fully — as they progress.   No one can put together a job history ten years after the client has moved in.

OTHER PROMOTION TOOLS

An ethical form of promotion open to professionals in the design field is the office open house.   This may take the form of art shows and other types of exhibits, seasonal parties, or hosting civic affairs.

Many offices regularly schedule exhibitions of staff paintings, sculpture, photography, and other art forms.   In medium- to large-sized cities local galleries are often open to the idea of lending works of art for temporary display.   The borrowing firm must provide adequate insurance coverage, of course.   One large architectural-engineering firm in Pennsylvania holds its annual employee art exhibit during the two-week period preceding Christmas, capping the event with the office Christmas party, to which several hundred local friends and clients are invited to join the staff and principals for refreshments and a tour of the office.

Certain civic groups are always interested in unusual meeting places and a designer's office, if he has a large conference room,

Building (formal name)_____

Owner _____

Location _____

Building cost (Total; per sq. ft.; or other) _____

Architect_____

Architect's address _____

_____

Personnel in architect's firm who should receive special credit (e.g., partner-in charge, project architect, designer, etc.)_____

_____

Associated architects (if any)_____

_____

Photographer (if publication quality photographs exist or are to be taken)

_____

Consultants to the architect (please give addresses):

    Structural Engineer_____

    _____

    Foundation or Soils Engineer _____

    _____

    Mechanical Engineer _____

    _____

    Electrical Engineer (if same as above, write "same") _____

    _____

Fig. 6.1

Acoustical Consultant _____

_____

Lighting Consultant (independent consultant only, not manufacturer)

_____

Interior Design Consultant _____

_____

Landscape Architect _____

_____

Graphics Consultant _____

_____

Cost Consultant _____

_____

Other specialist consultants (e.g., educational, library, stage, traffic, real estate, etc.) _____

_____

General Contractor _____

_____

Principal subcontractors: [Note: These firms will not necessarily be given credit in the article unless you specifically request it.]

Structural (fabricator and erector or specialist subcontractor)

_____

Mechanical _____

Electrical _____

Plumbing _____

Others who performed significant work _____

_____

Fig. 6.1 (Cont.)

117

is an ideal location. Boy Scouts and Explorer Scouts include architectural and engineering subjects in their programs. These young men will always welcome an opportunity to tour design offices and hear an explanation of the operation.

CAPITALIZE ON DESIGN AWARDS

Design awards of all types can be good promotional tools for the professional. Competitions are sponsored by professional organizations, magazines, suppliers, business development associations, and departments of state and federal governments. One must enter these competitions, obviously, before he can hope to win any awards. While a list of the winners is usually publicized, the losers are not, so the secret is to keep entering until a firm's work is recognized by an award. Representative competitions are those sponsored by:

American Association of Community Junior Colleges
American Association of Nurserymen
American Association of School Administrators
American Hospital Association
American Institute of Architects (National)
American Institute of Architects (Chapters, States, and Regions)
American Institute of Steel Construction
American Library Association
American Society of Church Architecture
Architectural Record (Interiors Awards Program)
Association of Medical Clinics
College & University Business Magazine
Design in Steel Awards Program (AISI)
National Conference on Religious Architecture
Prestressed Concrete Institute
Progressive Architecture Magazine Design Awards
Reynolds Aluminum Awards

Details about entry format and deadlines may be obtained from each of the sponsors.

EXPLAINING CLIENT SERVICES

Many architects and engineers seem to be unable to convey the usually understood range of their services in terms understandable to the layman client. This explanation is a key section of most brochures and usually is covered in presentations—but often badly. The two examples which follow differ considerably in format but both are aimed at explaining the role and responsibility of an architect in a manner almost any nonprofessional can grasp. The first is pure text; the second is a diagrammatic approach, shown in Figure 6-2.

SERVICES FOR CLIENTS—*The Role of the Architect*
1. Programming
Since the architect must eventually translate the client's ideas and needs —present and future—into working drawings and ultimately into a building, discussions in depth between architect and client must take place in the initial stages of project programming. One architect calls this phase "Reconciling dreams with dollars." From these early conferences emerge requirements and uses for the structure, along with the scope and essential characteristics of the project. The architect is alert to guide and assist the client in defining basic intent, purpose, and function, as well as space needs and relationships. In addition, evaluation and analysis of potential building sites and studies of traffic and utility service requirements are all directed toward establishing a timetable and proposed budget, and are important elements in the development of the overall design concept. This phase establishes the foundation or "program" on which the rest of the architect's work will be based. The program may be no more than voluminous notes and rough sketches, but if the project is a major one, a formal and detailed written program may result, usually prepared with the assistance and recommendations of expert consultants.
2. Master Planning and Preliminary Design Study
    (Schematic Design Phase)
With the building program established, the architect begins the translation of ideas into reality by preparing sketches, models, outline specifications, and preliminary cost estimates. The sketches include floor plans, site plans, and exterior views (elevations). Depending on the size and complexity of the project, as many as a dozen different study models may be built by staff designers and model makers to assist the client in making the necessary decisions at this stage. Obviously, this is a highly important phase of the whole procedure. Regarding cost estimates, all possible guesswork in this field should be eliminated, once the client's requirements are established in the preliminary conferences. Employing his own experience, along with data from many sources and the input of outside professional estimating firms, the architect is able to project substantially accurate estimates of the final cost of the project. Proven budget analysis and cost-control procedures are continually applied as the design proceeds through remaining stages.
3. Design Development
Sometimes rather confusingly referred to as "final preliminary design,"

this phase follows the client's approval of a schematic design from the various proposed solutions discussed above. Design development brings the earlier models and perspectives into sharper focus. The selected design undergoes thorough study as drawings are developed to determine more precisely planning, character, and construction. Major problems are defined and solutions worked out. The detailed preliminary design drawings and outline specifications prepared at this stage clearly illustrate and interpret the initial design concept in terms of siting; plan; form; character; materials; structural, mechanical, and electrical systems; and more detailed cost estimates.

4. Construction Drawings and Specifications

After the client has approved the design established in the "preliminaries," the "construction document" phase begins. This includes production of working drawings and the written specifications on which bids will be based and from which contractors will work.

Working drawings illustrate in detail all essential architectural, structural, plumbing, heating, cooling, electrical, and other mechanical systems and equipment. Elevations and sections and site, utility, and landscape plans are included in the working drawings.

The specifications reduce to writing all design elements not readily shown in the drawings and include materials to be used, design criteria, methods of installation, and prescribed levels of craftsmanship. Bidding forms and contract conditions are prepared and the project is ready for contractors' bids or negotiations with one or more construction firms.

5. Services during Construction

Following the selection of a contractor or contractors, the client is assisted in preparing contract documents, insurance certificates are secured and reviewed, and the scheduling of various contractors and subcontractors is coordinated. Shop drawings, material listings, and samples are checked and approved; change orders are prepared as required to insure that all changes from the original contract are a matter of record and handled in a businesslike manner; contractor's monthly statements are reviewed and certificates of payment are issued and monthly progress reports from the contractor are reviewed and submitted for the client's information.

Services during the building stage also include supervision of all phases of construction, maintaining current records of construction costs, and verifying that all work is performed in accordance with the requirements of the construction contract.

6. Total Design

Today's architectural offices, in increasing numbers, are prepared to assume responsibility for all elements of a client's building project. Harmonizing and unifying additional and supplemental services such as construction management, master planning, landscaping, graphics, computer services, and interior design are all available through consultants and in-house capabilities. Thus the client can look to one representative—his architect—to correlate every aspect of the design, construction, equipping, and furnishing of his project; from site selection to the placement of ashtrays in the finished building.

LECTURES AND SEMINARS

There are almost unlimited opportunities for public appearances open to the design professional. Unfortunately, many of these involve his talking to fellow professionals—a notoriously bad source of prospects. The trick here is to uncover lecture and other speaking possibilities where the audience may include some potential clients.

One example will suffice to illustrate the proper use of this promotional tool. A firm which has for many years counted hospital projects as almost 50 percent of its total practice regularly offers qualified speakers from its staff to university classes in hospital administration. The students in the audience are always at least in their fifth year of study and many soon will be going into hospital administration work with a master's degree.

The firm donates bound copies of past projects—master plans, programs, and small-scale copies of project plans—to the school's library. The guest lecturers provide each student and the professor with a large folder of material gathered from many sources, including, of course, the firm's own brochures.

The class discussion covers such subjects as planning, development, and construction, illustrated by the firm's motion picture prepared for showing to hospital and hospital-related prospects. Since this is usually the initial exposure of the students to a design firm—and their first opportunity to discuss some of the practical matters involved in renovation and new facilities construction, a friendly and interested reception is always assured the lecturers. Best of all, it is a chance to get the name of the design firm firmly implanted in the minds of up to forty future hospital administrators in a single presentation. The results of this type of contact do not manifest themselves overnight; indeed, five or ten years may pass before any one of these students is in a position to influence the selection of an architect or engineer. It must be viewed as a long-term investment of time and materials; but most design firms fully expect to be in operation five to ten years from today.

This discussion of promotional tools and strategy will conclude

BUILDING SEQUENCE

AND RESPONSIBILITY

		Owner	Architect	Joint
1.	Preliminary conferences			●
2.	Owner-architect agreement			●
3.	Establish building program			●
4.	Program analysis		●	
5.	Schematic designs		●	
6.	Approves schematic designs	●		
7.	Preliminary drawings		●	
8.	Preliminary estimates		●	
9.	Conference on preliminaries			●
10.	Revisions to preliminaries		●	
11.	Approves preliminary documents	●		
12.	Authorizes final documents	●		
13.	Final working drawings		●	
14.	Conference on specifics			●
15.	Final specifications		●	
16.	Final estimates		●	
17.	Conference and acceptance			●
18.	Approves final documents	●		
19.	Select contractors for bidding			●
20.	Issues documents for bidding		●	
21.	Receives bids	●		
22.	Bid tabulation and review			●
23.	Advises on contract award		●	
24.	Awards contract	●		
25.	Assists in execution of contract		●	
26.	Executes contract	●		
27.	Approves bonds and insurance		●	
28.	Supervises construction		●	
29.	Prepares field inspection reports		●	
30.	Reviews and approves shop drawings		●	
31.	Inspects and approves samples		●	
32.	Prepares monthly certificates		●	
33.	Pays construction costs monthly	●		
34.	Review construction reports			●
35.	Prepares and signs change orders		●	
36.	Countersigns change orders	●		
37.	Receives guarantees from contractor		●	
38.	Makes final inspection		●	
39.	Receives release of liens	●		
40.	Makes final payment	●		
41.	Accepts building	●		
42.	Celebration			●

Fig. 6.2 *Services for clients — graphic representation.*

with a story about how a firm once lost a job to a bound correspondence file.   A Midwestern-based manufacturer of pharma-
ceuticals was planning a major addition to one of its plants a few
years ago and interviewed a number of A-E offices in the process
of making a selection.   Finally, the choice was between two firms,
both of which had demonstrable experience with the building type
in question.

When one of the two offices was notified that the job had gone
to the other firm, the principal of the losing firm called the client
back to find out what had tipped the scales in favor of his competitor.   There were several reasons but the client seemed to emphasize one above the others.

The other office stressed their "design manual" in the presentation and in subsequent interviews with the client.   This design
manual was no more than a complete collection, in chronological
order, of copies of all correspondence, job-meeting minutes, change
orders, i.e., everything but the specs and working drawings, pertaining to a specific job.   Neatly bound in simulated leather, the
manual is presented to the client upon completion of his project
as a permanent record of everything written or typed during design and construction.

Every client gets the same material from his architect and engineer during the course of any job, but it comes in batches as
the work progresses, rather than in the neater book form.   The
winning firm had several samples of design manuals from past
projects to show prospective clients during presentations—and
the idea apparently worked for them.

We move on in the next two chapters to the key subject of presentations—the first plateau in getting the job.

CHAPTER 7

The Presentation: Prologue

In this chapter we will be concerned with surviving the selection process and with strategic and other considerations leading up to the formal confrontation—known as the interview—with the prospective client. No matter where or how information about a prospective job is turned up or how outstanding a professional's qualifications are to design it, if he doesn't get an interview and an opportunity to present his credentials, there is just no possibility that he will get the job. The next chapter covers strategy during and following the initial presentation and suggests some reactions to the potential client's final decision—whether it be good news or bad.

COMPLETE THE HOMEWORK

Before doing anything else, upon hearing about a potential job, complete the intelligence-gathering homework on the client and his project. Many of the business reference books listed at the end of Chapter 4 will be helpful at this stage. As a start, if the client is a corporation check out it and its executives in Dun & Bradstreet's *Reference Book of Corporate Managements* and Standard & Poor's *Register of Corporations, Directors and Executives*. It is often desirable to include information from *Who's Who in Finance and Industry* and *Who's Who in America,* both from Marquis Who's Who, Inc. If the top executives and directors do not appear in the larger *Who's Who in America,* try the appropriate regional edition of *Who's Who.*

If the client is a governmental body, find out the names of the top elected and appointed officials of the agency or department and look them up in Bowker's *Who's Who in American Politics.* On the federal level the *Congressional Directory,* the *U.S. Government Organization Manual,* and the *Congressional Staff Directory* will be helpful in varying degrees.

For private clients it is usually advisable to get the latest Dun & Bradstreet "Commercial Report." In addition to providing a quick rundown of the potential client's financial condition, the Commercial Report may suggest further leads on subsidiaries or parent companies and give additional information on names and the biographies of officers and directors. If there is a need for speed in obtaining the D&B report specify that it is a "priority" report. Priority requests should be answered by a telephone report in a day or two, followed by the usual printed report.

For corporate prospects there may well be an exploitative overlap in directorships; e.g., a director of a company for whom a firm has successfully completed projects in the past also sits on the board of the corporation the principals now want to interest in their services. Ties of any kind to the prospective client—schools, fraternities, relatives, past clients, hobbies, and friends—are all possibilities for intelligence gathering and bridge building. The *Who's Who* volumes are probably the best source of this type of incidental background information.

IS THE JOB REAL?

Concurrent with the development of general and specific knowledge about the prospect from various sources, one must also determine whether or not the project is apt to be built; confirm that it is a "real job," in other words. The state of the prospect's financing and the stage of his program development are useful inputs in making that determination. It is sometimes difficult to find out who is the real decision maker for a project, but little of substance can be accomplished in pursuing a job until it is known who will make the intermediate and final decisions.   The person being sought out is not necessarily the chairman of the board or the president or even the executive vice president of the corporation.   He may be a department head responsible for engineering or planning or new facilities.   His title and apparent function may bear no relation to his importance in this case.   In any event, the professional usually is on safe ground in cultivating the two or three top officers of the company and as many board members as he is able to contact, since they almost always will have an approval or veto power over the final selection.   During the search for the corporate decision maker it is wise also to learn the names of his immediate deputies and to meet them.   In the event of the principal's absence or—as happens—he resigns or is replaced during the selection process, there will still be a point of internal contact.

It may not seem very difficult to locate the power center for governmental projects: by his title, the top elected or appointed official is self-evident.   Making just such an assumption has caused many good design firms to waste days and even weeks of valuable time in cultivating mayors, city council presidents, city engineers, governors, and department heads on the federal level. While firm ABC is wining and dining the mayor, firm XYZ is entertaining the county political chairman or Mr. Doe, the party's major political contributor—and XYZ, everything else being equal, walks off with the commission.

It is seldom a simple matter to locate the decision maker in a project for a governmental body, especially in unfamiliar territory, but it is almost always the key to obtaining any meaningful consideration for a design firm.   Occasionally it will be discovered that a government job is "wired"; that it already has been promised to a firm and the selection process is being followed because it is required by law or custom or to make the deal look a little less phony.   Hopefully, a firm will learn of such charades before a lot of time has been spent on chasing the job.   As soon as the situation is confirmed to everyone's satisfaction the only recourse is to bow out gracefully and go on to the next prospect.   In my experience there is nothing to be gained by hollering foul or complaining. The burden of proof will be on the design professional and the conspirators will have left no evidence around on which any kind of case could be made, should one be inclined to try. Be philosophical —someday you may have an opportunity to go in wired.

In going into an unfamiliar city in quest of new business there are several sources of information available to anyone who takes the time to dig them out. First, go to the leading newspaper and take out a two- or three-month mail subscription. While at the paper talk to the financial and real estate editors. One can usually lay his cards on the table with these editors. The information they may give in return might only be a rehash of what has already appeared in the newspaper, but at least the designer will have that much more background than he went in with. These sessions must always be played by ear, but the opportunity may present itself to ask about the local movers and shakers in the community as well as in corporate and political affairs. A good rapport with the financial and real estate departments of hometown papers might move those editors to call or write ahead of a visit to provide the professional an introduction and some buildup with their out-of-town counterparts.

Do not overlook the broadcast media in a strange town. Television and radio news reporters also must know the local scene. In some cities television stations carry a locally slanted financial program. Talk to the people who put that show together.

And don't forget the chamber of commerce. If the town is upwards of 30,000 in population the chamber should be a prime source of information about local business, economic indicators, and identification of the decision makers.

A few other sources of intelligence information are the major local contractors or their AGC chapter, the urban redevelopment authority, city and county architect's and engineer's offices, local planning bodies, banks, and industrial and commercial development agencies—both public and private.

Obviously, past and present clients will not be overlooked as prime sources of information about a potential client, whenever and wherever a connection can be established. As the information begins to accumulate, the person in charge of business development should be able to complete the New Business Worksheet or Prospect Contact Form (as shown and discussed in Chapter 3).

VISIT CLIENT AND SITE

During the investigative stage it usually is a good idea to try to schedule a visit with the prospective client to confirm the scope of the job and to ask about certain other details on which more information is desired.   While this can be done just about as effectively by a telephone call, things are reaching the stage where it is important to establish a personal, face-to-face relationship with the potential client.   The trip may also be utilized for a site visit if its location is known.   Take a camera to the site and photograph it from several vantage points and from as many different directions as possible.   If they are available, obtain topo, utility, and political maps of the site and copies of boring test reports which have been made.   The site views and slides of some of the maps and studies can be used as part of the presentation to the client later on.   These graphics will also be helpful to the presentation team as background information about the project.   Besides, the potential client may be impressed by this attention to the details of his project.

Some of the site photographs shot on the preliminary visit may be found to be quite similar to views of other projects in an office. One firm made a photographic study of a long, narrow 105-acre site, bisected by a road, which was almost identical to a site for a corporate headquarters the design office had just completed a few miles away.   The photographer knew what pictures he had already of the completed building, so he duplicated views and angles as closely as possible on the undeveloped site, which, as it happened, was also for a corporate headquarters.   A special brochure was then prepared, showing the similarity of sites and the design firm's related, completed project, which also spanned a roadway, as one possible solution.   The corporate prospect seemed appreciative of this extra touch and the architect eventually realized some $60 million in work from the company, although many other factors besides the special brochure entered into obtaining the commissions.

By now enough information should have been assembled about the prospective job from a variety of dependable sources to enable the principals to decide whether or not they want to pursue it.   If this intensive intelligence-gathering stage goes against one's ethical or moral grain for some reason, be assured that the prospective client is having or will have every design professional he may be considering for his job thoroughly investigated—financially, ethically, philosophically, and morally.

It is possible to get carried away by this sort of thing, of course. A firm in Florida gave serious consideration to putting a private investigation agency on a retainer, to check out prospective clients as a matter of routine. Another gives preference to former CIA employees in hiring for its business development staff, on the premise that Agency men are best qualified by training and experience to develop the necessary intelligence on prospects. This firm may have a valid point, assuming that the exagent is also trained in architecture, engineering, planning, marketing, and promotion—and some are.

MAINTAIN CONTACT

If the decision is affirmative on going after a prospect, acknowledge interest promptly by a telephone call, followed by a letter. Support the letter of interest with brochures and other relevant material. This is the opening salvo—don't skimp or stint! If personal contact with the client has not yet been made through a site visit or other means, it is not a bad idea to hand deliver the material. Even if there already has been a meeting, deliver the letter and brochures in person if there is any reasonable excuse for doing so.

If a proposal is required as part of the submission, and the prospect's Request for Proposal (RFP) has been received, indicate in the letter of reply that the detailed proposal will follow promptly. Meet all deadlines for getting in letters of interest and proposals. It is not necessary, or even very desirable from the design professional's standpoint, to beat deadlines by large margins. If the RFP states that the proposal must be in the prospect's office by 5 P.M. on a certain date, many firms make it a practice to deliver it to him no earlier than 4:30 P.M.—the closer to the closing time the better. The rationale here is that the proposal will end up on top or near the top of the pile of proposals and therefore have greater visibility for the selection committee. Whether or not this technique has ever won a commission for a firm is open to question, but it is a common practice with proposal submissions.

ASK FOR HELP

Sending copies of the letter of interest with a brief note of explanation to past clients and business friends may generate some letters of reference and endorsement of a firm's services from these contacts to the potential client. Since this is now a critical stage of the selection process—and a firm's primary consideration is to survive cuts by the prospective client and to be scheduled for a formal presentation—it usually is better not to leave anything to chance. Depending on your relationship with past clients and business acquaintances, diplomatically—or not so diplomatically—let them understand that a call or letter from them to the prospect could be highly important to your chances of getting the job.

Prospective clients often ask for names of past clients and other references. Permission to furnish names of clients or other business references should always be requested, but this sometimes has to be after the fact. Naturally, names of disgruntled clients or creditors on the verge of suing will not be supplied. As was pointed out earlier, the practice of contacting past clients for reference purposes has been known to generate additional job leads and even definite projects—because they happened to be hit at just the right time. Design professionals should never lose contact with past clients and the request for a reference or endorsement is one of the many acceptable ways of keeping in touch.

A firm now has some peripheral assistance going for it in regard to selling the prospect on that firm. Do not neglect the prospective client at any point. There may be several occasions to meet with him while the sifting and cutting process goes on. Follow up every personal contact with notes to the appropriate parties. If your repertoire has not already been exhausted, enclose copies of previously unsent but relevant reprints with the notes.

This is also the time to update and complete the intelligence dossier on the potential client, his business, civic and fraternal organizations, friends, military service, children, hobbies, etc. By finding several areas of common interest it is much easier to develop the necessary strong personal relationships with the client and his representatives. As part of the total concept of "building bridges" to the prospect, always be aware of the potential of mature, responsible political and business influences which may be brought to bear at this point. In practical politics this is known as "clout."   In selling professional services the pressures should be less obvious, always ethical, and in a lower and less strident key than in politics, but pressure or clout it is—and those who know how and when to apply it leave interviews with the job more often than not.

PROFESSIONAL
AUDIOVISUAL PRESENTATIONS

For better or for worse, most design professionals build their presentations around the audiovisual approach.   Since this is the case and recognizing that most, if not all, of the presentations to a potential client will undoubtedly utilize projectors, slides, film, and a screen, today's design firm must be familiar with the latest and best audiovisual techniques and equipment.   These include slides, motion pictures, closed circuit television, flip charts, large display boards, and mixed-media shows.   Anyone in a firm who works with presentations should be a regular reader of *Audio-Visual Communications*, a controlled-circulation magazine published by United Business Publications, 200 Madison Avenue, New York, N.Y. 10016.   Write to the circulation manager on your business letterhead to determine whether or not you qualify for a complimentary subscription.

PLAN FOR THE UNEXPECTED

"Contingency planning" has been a much-heard phrase in recent years, following the publication of the *Pentagon Papers* and other examples of advance planning by government and industry.   Not to prepare for contingencies in presentation planning is to invite disaster.

Professor George T. Vardaman, who covers the theory of presentations in his book, *Effective Communication of Ideas*, has this to say about contingency planning:

> Any experienced speaker will tell you that, even with the best planning and execution of a presentation, the unexpected will happen.   We are, after all, human beings, not God.   So you'll be well advised to "expect the unexpected" (and if nothing untoward happens, great!).
>
> But even here you can anticipate some outcomes which are more likely than others.   For one thing your analysis of primary *receiver roles*

can be off, so you should have in mind the most likely initial-backup-desired contingency.   And this, of course, leads to a revision of your *outline of ideas*, meaning that you'll need a backup plan to accompany your contingent receiver role handling.   Backup for *equipment and aids* is also very important.   What if you have a power shortage?   Could you use the blackboard to sketch your flow chart?   What if the projector lamp burns out in the middle of a film showing?   Do you carry a spare and do you know how to put it in?   What if your planned handouts cannot be printed on time?   Do you have some alternate devices (e.g., a mimeographed outline) which would suffice?   What about last minute changes of schedule (e.g., you find that the preceding speaker went 30 minutes overtime, forcing you to collapse your planned one hour presentation to 30 minutes)?   These program changes happen more often than realized, and you'd best be prepared to meet them by having an "abbreviated" or "elongated" presentation in mind.

There is no need to try to list all the many unpredictables.   If you carefully prepare to handle those we've mentioned, and then deliberately assume that your presentation will probably never go completely according to plan, you'll be able to effectively handle almost anything that arises.[1]

MURPHY'S LAW

In his last sentence, Professor Vardaman infers the results of the operation of Murphy's law on the behavior of inanimate objects. Murphy's law, usually stated, "If anything can go wrong, it will," has been documented and refined by no less a professional group than the Institute of Electrical and Electronic Engineers.   D. L. Klipstein, in an article in the August 1967 issue of *Electrical and Electronic Engineers*, formulated Murphy's law in mathematical terms:

$$1 + 1 \, \text{☞} \, 2$$

where ☞ is the mathematical symbol for "hardly ever equals."

Frank R. Harris, public relations counsel, gives a few examples of Murphy's law in operation, with which most readers should quickly and painfully relate (some of the examples have been slightly adapted to the subject of selling professional services):

1. Following delivery of photographs to be included in an important brochure, which required chartering a plane to southern Mexico, the managing partner decides he likes the vacation snapshots taken by his wife better.

2. Three applications of Murphy's law universally come into play at dedications:

    a. Either no one will come or everyone will come.

    b. The public address system will not operate.

    c. It will rain.

  3. For presentations:

    a. The importance of the potential client will be in inverse ratio to the welcome he receives from the receptionist.

    b. When the principal of the firm visits the men's room ten minutes before an in-house presentation on a $25 million project is scheduled, his zipper will not close.

    c. When presenting in the client's boardroom, his building's power system will fail 7.25 minutes after starting.

    d. Time allotted to each person on a three-man presentation panel will invariably break down in practice in these proportions:

First speaker—67.0 percent of the time

Second speaker—32.8 percent of the time

Third speaker—0.2 percent of the time

    e. After the presentation, the managing partner will remember at least one vital statement he forgot to make.[2]

Several variations on Murphy's original law have appeared in recent years:

> Left to themselves, things always go from bad to worse.
>
> If there is a possibility of several things going wrong, the one that will go wrong is the one that will do the most damage.
>
> Nature always sides with the hidden flaw.
>
> If everything seems to be going well, you have obviously overlooked something.

CHECK OUT EVERYTHING

Let's assume, for the moment, that the formal presentation will be held in the client's office instead of in the designer's conference room. Try to visit the room where the interviews are to be held prior to the presentation date. Not long ago an architect discovered that a county client in Virginia planned to hold afternoon interviews in a conference room on the west side of the new courthouse. Unfortunately, there were no provisions for darkening the room and no other space was available for the interviews. His solution was to rent a rear projection box designed to accommodate the Kodak Carousel 35-mm projector. Radiant's "Univision" is one model of the rear projection box, featuring a large screen easily viewed under daylight conditions. Apparently, none of the other firms presenting to the county's selection committee took the trouble to check out the room ahead of time. One can imagine the chagrin of the other principals as they discovered their slides were all but invisible in the sun-filled conference room.

In addition to planning on how to cope with emergencies during a presentation, every piece of equipment which might be used should be thoroughly checked out well in advance of its use. This includes, but is not limited to, still and motion picture projectors; slides, films, and reels; screens; wall charts; and even the lights in the conference room. Projector lenses should be carefully cleaned before each use. Projector lamps are usually rated for so many hours of use. Some firms make it a point to change bulbs at the end of two-thirds or three-fourths of their rated life, to avoid burn-outs in the middle of an important presentation. Polish the reflectors behind the projection bulbs occasionally. Power and extension cords for projectors are among the world's most mislaid items —be sure there is a full set on hand at the same time the check is made to be certain that there is an extra projection bulb in the carrying case.

Motion picture film can get dirty, scratched, broken, and rewound backwards or even upside down by inexperienced operators. Slides of all sizes are subject to various problems, including dirt, fingerprints, and broken cover glasses in the heavier duty mounts. Ideally, all films and slides are inspected for damage following each use, before being returned to storage. The operative term here is "preventive maintenance."

Slide storage should never be a hit-or-miss affair. The ideal components of a functional storage system for slides include:

- Quick review of large collections
- Easy selectivity
- Safe, clean, catalogued storage
- Adaptability to various sizes and mounts

If the system includes its own viewing table or lighted upright screen, so much the better. Among others on the market, the ABODIA slide storage system, made in West Germany and distributed in the United States by Eldon Enterprises of Charleston, West Virginia, meets the above criteria. ABODIA units can be obtained to handle and view from 1,000 to 10,000 individual slides.

The experienced professional never uses audiovisual equipment without checking it out. Dry runs are held for slide shows and motion picture showings. Slides are checked for cleanliness, un-damaged mounts, and proper placement in the slide holder. One really doesn't need the distraction of a 35-mm slide featuring a large clear thumbprint in the center of the projected view or back-ward-reading legends on site plans and elevations. To many clients an upside-down slide is the mark of an uncaring, unpre-pared, sloppy amateur—and what design professional needs that kind of reputation when he is trying to sell his firm?

PRESENTATION DRY RUNS

Dry runs for the A-V equipment bring up the collateral subject of dry runs or "murder sessions" for the entire presentation and all those who will participate in them. Murder sessions, also known as "murder boards" around the Pentagon, involve deadly serious role-playing sessions used routinely by industry and departments and agencies of the federal government in preparing for important presentations to clients and congressional committees.

More and more architects, engineers, and planners are taking the time to hold one or more dry runs preceding the actual presenta-tions, with staff members taking the part of the potential client. (For murder-board sessions at the Pentagon, officers impersonate congressmen and toss tough questions in rapid-fire order at their fellow officers who are scheduled to testify.) All of the hard questions should be asked, with associates adopting a skeptical, even hostile, role at times.

SAMPLE "MURDER SESSION"

In covering this subject for a seminar on business development held not long ago all attendees were given the following set of "facts" about an approaching interview.

GIVEN

You are with the firm of Doe, Roe & Smith, Architects, founded in 1954 by John Doe. Roe and Smith became partners in 1960. You have some mechanical and electrical engineering capability in-house. Con-sultants are used for other engineering disciplines.

Your total staff, including engineers and principals, numbers thirty-seven. In the seventeen years of its existence the firm has developed a

fair practice mix, including schools, office buildings for both governmental and private clients, a complete small community college, some
planning work, and a few industrial buildings. Over the past five years
billings have shown an average annual growth of 15 percent. The firm
has won local and state AIA design awards; in 1968 you won a national
AIA Honor Award for a small research center for IBM. Billings for
1970 were $1,300,000.

The prospective client is Continental Mutual Assurance Company,
a medium-sized publicly owned insurance company with corporate
headquarters in New York City. For a variety of reasons the CMA
board of directors and company officials have decided to move their
operations out of New York to a suburban location near Elizabeth, New
Jersey. Several sites have been quietly optioned by real estate agents
acting for CMA and these locations all are under active consideration.

The company hopes to make the move in three stages, with approximately 800 employees involved in the initial move. This first group will
consist of primarily middle management and office personnel. In the
second phase an additional 200 persons will be moved, many of whom
are involved in computer operations. The final move will bring out the
remaining management and sales staff—about 100 persons altogether.

Continental Mutual Assurance has been in its New York offices since
1948. Three regional offices have been built by the company in the
past fifteen years; the most recently completed one opened in St. Louis
last year. Local architects were retained to design the regional offices
(all of which are in cities of over 500,000 population), except the one in
Miami. One of the firms which are in competition for the New Jersey
structure designed CMS's Miami office.

There are nine men on CMA's board of directors. One, the president
of a St. Louis bank, is also a director of Monsanto. Another director is
a member of Princeton University's board of trustees. The chairman of
the board is a retired admiral who commanded the navy's antisubmarine
patrol in the Atlantic during World War II. The executive vice president
of CMA (also a director) at one time taught economics at the University
of Missouri. This official is also active in Democratic politics in New
York State.

You have learned that CMA is seriously considering five offices, including Doe, Roe & Smith. Except for the firm that designed the Miami
regional office, you do not know any of the other four firms under consideration. Yours will be the third office to be interviewed. Two firms
were visited last week and the selection committee is coming to your city
from St. Louis, where they interviewed one of the firms yesterday.

Total cost of the project could reach $20 million, including land costs.
CMA wants its architect, following the selection, to do a complete investigation and evaluation of the four sites now under option. One site of
eighty-five acres lies along the New Jersey Turnpike, while the others
vary in distance from three to twelve miles from the turnpike.

The selection committee is composed of the executive vice president, the chairmen of the finance committee (the board member who is also on Princeton's board of trustees), and the vice president in charge of development and investment for CMA.

You have already greeted the selection committee as they arrived in your lobby, shown them John Doe's office on the way to the main conference room, seen that everyone had coffee, and have just finished showing them a selection of representative slides of your work, including:

Office buildings—high rise and campus plan
Several planning jobs
Industrial buildings, most of which included office space
The IBM research center

The lights have just been turned back on in the conference room and the discussion commences. As Doe, Roe & Smith's representatives you have forty-five minutes remaining for the interview.

Two of the seminar participants were selected to act out the roles of principals Roe and Smith; the remainder of the group was given copies of the selection committee's list of questions for its architects. The men taking the parts of Roe and Smith had until the next morning to prepare their presentation, based on the information given above. Some of the questions, which follow, might sound a little strange, but you have my guarantee that each of them has been asked at least once during actual interviews over the past few years.

QUESTIONS

How large did you say your firm is?
How do you handle engineering and other consultants?
Have you ever used XYZ & Associates in site investigations?
What do you think of their work?
Who would you prefer as a soils and foundation consultant? Why?
Will being alongside the New Jersey Turnpike be an advantage or a disadvantage, in your opinion?
What are some examples of other similar site studies Doe, Roe & Smith have done for clients? Have you ever had to recommend against all preselected sites and—in effect—start all over again?
Where is Mr. Doe?
If you are retained as our architects will Mr. Doe play any part in the design of the building? Who will be in day-to-day charge of the job—or does that vary depending on what stage the job is in?
How many square feet and what costs are we probably talking about in the first phase—for the first 800 people?
Have you designed many computer room setups?
It's only fair to point out that we've made some investigation of your

past work, as we have done with the other offices under consideration, and some of your past clients seem to feel that your firm is strongly design-oriented—perhaps too much so. Have you ever heard any remarks along that line, or would you care to comment on it?

(Directed to Roe or Smith)   How long have you been with Doe, Roe & Smith?

Why is it that you have only mechanical-electrical engineering in-house? It seems that structural engineers would be just as important to you and your clients.

Who are some of the consulting firms with whom you have worked?

When could you begin work on our project, if we made our choice by the end of the month and you are the firm we decide on?

What do you consider as the major concerns of an office building in an urban area such as central New Jersey?

Are you familiar with the concept of office landscape? Have you used it for any clients?

We saw some fairly large jobs in the slides you just showed and we know that Doe, Roe & Smith is pretty busy—could you really do justice to a project of the size we are talking about?

Other than Mr. Doe, what are some representative salaries of your top executives?

How have your early estimates of construction costs held up in relation to actual costs of the finished projects? What is your approach to estimating?

We will have to obtain a change in the zoning classification for our site. What has your experience been in taking projects through zoning hearings?

How much would your fee be for a project like ours?

Do you work only on a percentage of construction costs?

You say you can either work on T&M or a negotiated fee; doesn't T&M usually save the client some money?

What is your projection of the schedule, if we make a decision in two weeks to retain your firm?

1. Preliminary design
2. Construction drawings
3. Construction
4. Completion

What contractors, in your opinion, are particularly good in a negotiated fee situation?

What is the dollar value of work on your boards as of today?

Are you currently involved in any litigation with clients? If so, could you give us some details?

What are some of your past and current jobs you would suggest we visit in connection with making our selection? (Extra credit on this if one of the Doe, Roe & Smith representatives suggests that they set up appointments and accompany the CMA selection committee.)

What are three reasons you can give us to carry back to our board to justify selecting Doe, Roe & Smith over the four other architectural offices — all of which enjoy good reputations among their clients and their fellow professionals?

The Doe, Roe & Smith representatives *should* have opened this discussion by relating the interview to CMA's problems, rather than to problems of other clients.   The Doe, Roe & Smith people should not restrict themselves solely to responding to the above questions but should amplify each other's answers and volunteer information and comments throughout.   Occasionally, the discussion should veer away from architecture and the client representatives drawn out on other subjects and interests.

VIDEO-TAPE TRAINING

An office might use something similar to the imaginary Continental Mutual Assurance client background and the questions developed for Doe, Roe & Smith to establish the idea of holding murder sessions among its own principals.   Eventually the technique should be applied to actual client presentations.   As was pointed out earlier, some firms record mock presentations on video-tape for later playback and review over a closed-circuit television system. This same internal television system (camera and playback unit), with a few additional pieces of equipment such as a control console, a screen splitter, and a special effects generator can be used for very effective in-house presentations.

The installation of a closed-circuit television system is not an automatic guarantee of better presentations or improved internal communications.   Personnel must be trained to operate the system and the principals must learn to use it to the fullest advantage.

On this subject, Roger Ailes, former producer of the Mike Douglas television show, communications consultant to President Nixon during the 1967 election campaign, and now head of the New York radio-television consulting and public relations firm, Roger Ailes, Inc., had this to say in a recent interview:

> I disagree with many of the education-oriented theories.   They're so much B.S.   Ours is a video-oriented society and the man who sits down to watch a videotape . . . program has been conditioned by hours of watching broadcast television.   He expects action-oriented, high standard productions.
>
> Remember, the persons viewing the tape are not making a distinction between closed-circuit and broadcast television in their expecta-

tions. . . . Use show business techniques and make the presentation in an entertaining fashion. . . . Build interest. Capture the audience.[3]

Whether or not there are plans to incorporate closed-circuit television into a firm's own communications system, remember that more and more clients are turning to such facilities and design professionals should be conversant with current techniques and equipment. The quarterly *Wilke Report*, published for architects and interior designers by Hubert Wilke, Inc., Communications Facilities Consultants, 347 Madison Avenue, New York, N.Y. 10017, can be helpful with trends and the latest major A-V installations. A regular column, "The Drawing Board," discusses such subjects as "What's Happened to Projection Ports?" (Summer 1971 issue). Write Hubert Wilke on your letterhead for a copy of his interesting newsletter.

THE PLACE FOR MULTIMEDIA

One of the glamor words in today's audiovisual lexicon is "multimedia." The term usually is more properly "multiscreen," in that only one medium—film, both still and motion picture—is employed, but to fill a number of screens, either simultaneously or intermittently. A few of the larger design firms have spent many thousands of dollars to develop six-, nine-, and even twelve-screen productions, with the attendant hardware. An article in *Audio-Visual Communications*, "Multi-Media: Methods and Myths," pointed out some of the pitfalls of the technique:

> Cramming a multi-media program with too many details tends to decrease rather than increase retention. . . . If the objective is to impart detailed information, then a more straight-forward presentation might do as well with less expense.
> Another error is to put so much attention into hardware that software gets the short end of the stick. The hardware itself, the mechanics of the production, can be fascinating to the audience. . . . Try to keep the hardware hidden so it doesn't distract attention from the "message."[4]

The best approach to an A-V system is to shun overly complicated setups and avoid A-V overkill—wherein the audience is overpowered but not informed. The more pieces of complicated equipment, the greater the chance of a malfunction. Also, the total gross weight and logistics involved in transporting a multiple-projector, multiple-screen system, plus control centers and wiring, are things to consider when it is a one- or two-man presentation in the client's office. "Keep it simple" is a cardinal rule for audio-visual presentations—particularly when they are scheduled outside a firm's own quarters.

MULTISCREEN SHORT CUTS

Much the same effect as that obtained from a multiple projector and screen installation can often be achieved by a single projector and one large screen. Ernest Burden, writing on "Slide Presentations" in *Architectural Record,* described "multiple projection made easy."

> . . . competition rules dictated [the] form [of the slide presentation "Facade"]. The subject matter was selected for its timeliness in the ever-increasing war on community ugliness. It depicted the story of a group of buildings slated for destruction and the eventual construction of new buildings to take their place. To portray this amount of material on the screen would have required multiple projection, but the competition rules did not allow this. Therefore, all the photographs selected were prearranged, cut out and mounted on black cardboard. Then the composite was rephotographed onto one single 35-mm frame. This allowed up to 250 pictures to be shown on 80 single 35-mm slides. Certain parts of the slides were carefully masked to crop off any area not wanted on the composite slide. The end result was a multiple-image presentation using a single projector and a single slide each time. The possibilities of combinations using this approach are endless.[5]

Some years ago Edward Durell Stone used an adaptation of the technique described by Burden. In that case two Kodak 35-mm slide projectors were connected by Kodak's Carousel Dissolve Control to show progressive views of Stone's U.S. Embassy at New Delhi and the U.S. Pavillion at the Brussels World's Fair, as shown in Figure 7-1.

On another Stone project, the Fort Worth, Texas, City Hall, much the same setup was used to show the three-phased development of the city building. With the lap-dissolve effect the structure appeared to grow effortlessly from a one-story into a three-story building. A good model and painstaking model photography are the keys to a smooth transition of the views.

The Eastman Kodak Company publishes a wealth of material for the audiovisual enthusiast. Introduce yourself to this storehouse of information by obtaining a copy of the "Kodak Publication List (Audiovisual and Motion Picture)" from your camera dealer. If he doesn't have it, write to the Motion Picture and Education

Fig. 7.1 *Examples of split-screen 35-mm presentation slides.   U.S. Embassy in New Delhi. (Courtesy Edward Durell Stone and Associates.)*

142

Fig. 7.1 (Cont.) *U.S. Pavilion at the Brussels World's Fair. (Courtesy Edward Durell Stone and Associates.)*

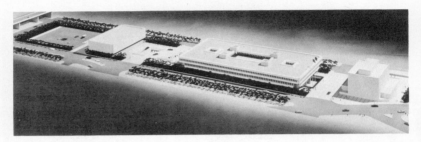

Fig. 7.1 (Cont.) *Fort Worth City Hall development. (Courtesy Edward Durell Stone and Associates.)*

Markets Division, Eastman Kodak Company, Rochester, New York 14650. "Slides with a Purpose," No. V1-15, for 20 cents, is a good introductory pamphlet from Kodak on the subject, but there are many other excellent technical pamphlets and books available.

REFERENCES

[1] George T. Vardaman, *Effective Communication of Ideas*, Van Nostrand Reinhold Co., New York, 1970, p. 137.
[2] Frank R. Harris, "If Anything Can Go Wrong—It Will," *Public Relations Journal*, New York, February 1972, pp. 40–42.

[3] "The Communicators: An Interview with Roger Ailes," *Audio-Visual Communications*, September 1971, pp. 14ff.

[4] "Multi-Media: Methods and Myths," *Audio-Visual Communications*, June 1971, pp. 14–16.

[5] Ernest Burden, "Slide Presentations," *Architectural Record*, July 1971, p. 57.

CHAPTER 8

The Presentation: The Interview

The interview is the payoff for all that has been done, said, and written up to now. Reaching the interview stage is the first plateau—the semifinals in the business development sweepstakes. A winning presentation has the proper mixture of information, empathy, ego, and good "show biz." Most of the factors which separate the highly successful firms from those who just exist on a hand-to-mouth basis are displayed during a presentation. The successful principal, representing his firm in an interview, knows how and when to "go for the jugular"—and watching a master salesman-designer, who has done all of his homework, playing a prospective client during a presentation can be a wondrous thing— and a real education.

Warren Leslie, president of a New York public relations firm, put it, ". . . there is some theater in all business and a great deal in the volatile world of advertising and promotion. It shows itself in presentations, in meetings, in all kinds of selling, and the man who hasn't some sense of theater in him will stand still. Theater means timing and force."[1]

TRIM = EFFECTIVE COMMUNICATIONS

Professor Vardaman, in his previously cited work on effective communications, reduces the basic ingredients of the procedure to the acronym TRIM:

T— the communication target at which you are shooting.
R— the receiver to which your communication is addressed.
I— the impact needed to influence the receiver as desired.
M—the methods which can be used to get the job done.[2]

The Target and the Receiver, Professor Vardaman's two initial TRIM elements, usually are pretty well self-defined for an architectural or engineering presentation. The impact (I) required— how the presentation must affect the receiver (potential client) if you are to hit the target and influence him as needed—has been the underlying subject of most of the preceding chapters. This chapter is concerned primarily with methods (M), the presentation payoff. On this point Professor Vardaman urges that the presenter carefully specify the presentation target, the presentation receiver, the impacts and devices to be utilized, and, finally, outline ideas at the outset of his presentation planning.[3]

The ideal presentation, according to the principals of a number and variety of firms, would:

1. Be the last interview scheduled by the prospective client
2. Be held in the design professional's office
3. Consist of a formal, structured presentation, followed by a tour of the office, drafting rooms, and service and other support areas
4. Include lunch with the potential client in an informal, quiet, unhurried atmosphere
5. Conclude with a tour of some of the firm's completed work related to the prospect's project

AVOID MASS INTERVIEWS

A recurring horror story in the design professions concerns the potential client who sets out to interview forty or so firms for his project. All too often the project turns out to be a $200,000 school addition or a $500,000 hospital renovation. Seldom is the client aware of the imposition he makes on architects and engineers in such cases—or how utterly useless and confusing the mass interviewing procedure can be for all parties involved.

The experienced firm does not respond to these invitations when their nature can be determined in advance. But even in the more normal interview situation, where the client has cut his list of potential firms to a reasonable and manageable number of, say, three to five offices, the principals of many firms believe there is an advantage to being the last scheduled interviewee. Sometimes the time span to be covered by the interviews can be determined in advance and a request made that your firm be scheduled for the last day. In other cases, a slightly more abstruse route must be taken to obtain the ultimate or penultimate interview. Any supposed psychological advantages of being the finalist in a series of interviews is based on the theory (or hunch) that the potential client is more apt to retain the important elements of the last presentation he hears.

The opposing school on interview placement prefers to be scheduled as early as possible, on the premise that the members of a selection committee will come fresh to their task and are more apt to fully assimilate and remember the fine points of a designer's presentation.

Whether or not these are sound psychological bases to bolster either side's stand on the best time to present, there are a number of valid reasons for trying to schedule the formal presentation in the professional's office, rather than on the home ground of the potential client. Perhaps the primary advantage is that the firm is enabled to exercise complete control over all arrangements and elements of the visit and interview. It also allows the full VIP treatment to be extended to the client, particularly if he is from out of town.

VIP TREATMENT

On this subject—VIP treatment—there are a few standard techniques and considerations which might be well to review here.

If an overnight stay is involved, offer to make hotel reservations. If there is a well-known private club in the city, a room there is often preferable to commercial hotel accommodations. Make sure the client is issued a guest card for the club's facilities.

Arrange to have the client met at the train station or airport, preferably by one of the firm's principals who has already had contact with him. Use judgment and basic public relations techniques in such meetings; one just doesn't pick up an official of the Ford Motor Company in a Cadillac, for example.

Should the client not be from out of town or should he prefer to make his own way to the designer's office from the airport or train depot, make very sure there will be a welcoming committee of top management awaiting him in the lobby.

Set up all coffee breaks, luncheon and dinner arrangements, and menus well in advance. If the meeting in the firm's offices gets underway before 10 A.M., it is a good idea to offer the visitor coffee and rolls on his arrival.

The *out-of-office visits to completed work in the area* can be a very important part of the whole selling process, if handled properly and professionally. Try to select projects which are related to the client's and which exemplify the type of design approach it is believed the client wants for his building. Be sure a client representative will be present at each project to be visited. These people can say things about a professional's work which would sound immodest or unethical coming from the designer. (By the same token, the client representative—if not carefully selected and briefed in advance—can bring all hopes for the new commission to an ignominious end by giving the wrong answers or by volunteering unwelcome details about roof leaks, omitted storage closets, and similar items.)

An important point to keep in mind on these project tours is never to talk disparagingly about the work of other designers which may be seen, particularly if the potential client asks direct questions about a building not designed by his host firm. A negative reaction would demonstrate an unethical attitude, of course, but more important, the potential client may know and admire the other professional.

Include an offer of entertainment following dinner if the client will be in town overnight; the theater, a night club, or a small party in your home are possibilities. If the client is from a large city with many legitimate theaters he will probably enjoy a relaxing evening in the home of one of the principals much more than being dragged off to a third-rate hometown play production. If he is not a swinger type, don't suggest a nightclub tour, either.

If the visit of the potential client will occupy most of one day or even run over into the following day, *a typed-out schedule of the*

arrangements made is usually a good idea.   Give the visitor a copy
of his schedule, edited if necessary.   This might sound a bit too
formal and overly structured, but it is always better to err on the
side of overplanning.   A multimillion dollar project could be at
stake.

PRINCELY TREATMENT

For those readers who have never had an opportunity or the necessity to become involved in high-level VIP scheduling, the following
exhibit may be of interest.   It details only the press arrangements
set up for the 21½-hour visit of Prince Philip to New York City in
November 1969, and was made available through the courtesy of
Eric N. Smith, British Information Services, New York, N.Y.

— — — — — — — — — — —

Note to Editors

PRESS ARRANGEMENTS FOR VISIT OF
HIS ROYAL HIGHNESS THE PRINCE PHILIP, DUKE OF EDINBURGH
TO NEW YORK
NOVEMBER 5–6, 1969

Press Facilities

Details of the program and of the press facilities and contacts for the
various events are attached.

Press Credentials

There will be no special accreditation.   Current New York Police Working Press Cards (or similar passes from other cities) will be sufficient
identification.

Enquiries

For enquiries about the visit to New York please contact:—

British Infomation Services   Mr. Eric N. Smith
845 Third Avenue   Extension 376
New York, N.Y. 10022   or
Telephone: 752-8400   Mr. Ian Brett
                                     Extension 419

or the contacts indicated against specific events listed on the attached
program.

VISIT OF
HIS ROYAL HIGHNESS THE PRINCE PHILIP,
DUKE OF EDINBURGH

TIME (All timings approximate)	OFFICIAL EVENTS	PRESS ARRANGEMENTS AND CONTACTS
WEDNESDAY *November 5*		
1:00 P.M.	His Royal Highness arrives at the Marine Air Terminal, La Guardia Airport, accompanied by Her Majesty's Ambassador and Mrs. Freeman. Met by representatives of the Governor of New York and Mayor of New York City and by the British Consul-General and Lady Rouse.	Press Conference at Gate No. 1, Marine Air Terminal, La Guardia. (Arrangements will be the same as for a Presidential arrival.)
1:15 P.M.	Departure from La Guardia Airport.	
1:45 P.M.	His Royal Highness arrives at Madison Square Garden for lunch with the President of the National Horse Show Association of America, Mr. Walter B. Devereux.	Photography only—on arrival outside Madison Square Garden at 4 Pennsylvania Plaza entrance (31st Street, between 7th and 8th Avenues).
3:00 P.M. (Approximately)	His Royal Highness attends the matinee performance at the National Horse Show.	Facility for photographers, including TV, at a time to be decided between 3 P.M. and 4 P.M. (no interviews). A light buffet lunch will be available in the Press Lounge on the 1st Promenade level (entrance from 8 Pennsylvania Plaza, 33rd Street, between 7th and 8th Avenues) from 2:00 P.M. onwards. *Contacts* Michael Sean O'Shea, Marianne Strong Assocs., PR for National Horse Show Association of America Telephone: 249-1000 (Mr. O'Shea can be con-

VISIT OF
HIS ROYAL HIGHNESS THE PRINCE PHILIP,
DUKE OF EDINBURGH

TIME (All timings approximate)	OFFICIAL EVENTS	PRESS ARRANGEMENTS AND CONTACTS
		tacted on November 5 at Madison Square Garden— Telephone: 594-6600.) For advance information relative to Madison Square Garden call—Mr. Bill Shannon, PR for Madison Square Garden—Telephone: 594-6600.
4:30 P.M.	Depart Madison Square Garden.	
5:00 P.M.	Arrive at the Headquarters of the English-speaking Union, 16 East 69th Street.	Photography only—on arrival outside the E.S.U. Headquarters building. *Contact* Mrs. E. Sternberg English-Speaking Union. Telephone: TR 9-6800
6:00 P.M.	Depart E.S.U. for Waldorf Towers.	
7:00 P.M.	The Secretary-General of the United Nations calls on His Royal Highness at the Waldorf Towers.	No press or photograph facilities. Official United Nations photographs will be available later the same evening at the U.N. Secretariat building. *Contact* Mr. Marvin Weill, Office of Public Information, United Nations Secretariat. Telephone: PL 4-1234 Ext. 2863
7:30 P.M.	His Royal Highness meets Gov. Nelson Rockefeller and the Executive Committee of the Pilgrims of the United States.	7:30 P.M.: Facility for photographers, including TV, in the Crane Suite on the 4th Floor of the Waldorf-Astoria Hotel. (Use Park Avenue entrance and take West elevators.) *Contacts* Mr. Hugh Bullock President of the Pilgrims of the United States.

TIME (All timings approximate)	OFFICIAL EVENTS	PRESS ARRANGEMENTS AND CONTACTS
		Telephone: BO 9-8800 Or 943-0635 For information relative to the Waldorf-Astoria Hotel contact: Miss Lola Preiss, Waldorf-Astoria Hotel. Telephone: EL 5-3000.
8:00 P.M.	His Royal Highness dines with the Pilgrims of the United States in the Star- light Roof on the 18th Floor of the Waldorf- Astoria Hotel.	8:30 P.M.: Refreshments will be available for the press in the Gold Room on the 18th Floor. (Use Park Avenue entrance and take West elevators.) Speeches by Governor Rockefeller, Mr. Bullock and Prince Philip (sound only) will later be relayed from the Starlight roof to the Gold Room from about 9:15 P.M. onwards. A feed line will also be available in the Gold Room for those who wish to hook up their own microphones. Contact should be made *in advance* with: Miss Lola Preiss Waldorf-Astoria Telephone: EL 5-3000
THURSDAY *November 6* 8:30 A.M.	His Royal Highness leaves the Waldorf Towers for the Marine Air Terminal, La Guardia.	
9:00 A.M. (Approxi- mately)	Arrival at Marine Air Ter- minal, La Guardia.	Photograph facilities only at Gate No. 1, Marine Air Terminal, La Guardia.
9:30 A.M.	Departure for United Kingdom.	

Naturally, there is no suggestion that a schedule for a potential client visit should be this detailed (unless a royal client is involved), but Prince Philip's 21½-hour New York City itinerary might stimulate a few ideas for the care and feeding of future clients.   Above all, do not be afraid to lay it on.   The odds are that the selection committee has at least one senior officer of the company among its members—and he will be accustomed to kid-glove, red-carpet handling by his own staff.   If your firm takes care of such details in a professional and sensitive manner, while the other candidate offices do only a mediocre job or overlook the niceties altogether, this will be one more point in favor of the firm when the selection committee sits down to decide on its designer.

PRESENTATION GUIDELINES

Since there are an infinite number of variables in every interview; i.e., no combination of personalities and backgrounds of a potential client, the requirements and functions of the project, and the experience and approach of the design firm interviewee will ever be exactly the same, there is little point in going through the exercise of outlining specific steps for a "model interview."   On the other hand, a few suggestions and helpful hints can be set down for general guidance.   These should be incorporated into and adapted to an interview-presentation only when indicated by circumstances and on the basis of good judgment.

As a *general* rule, do not begin the interview with a slide presentation.   Use the opening minutes of the allotted time to establish rapport between client and the design professional's representatives.   Explain to the client that some pictures of the firm's past and current work will be shown and discussed with him, but since his primary interest is in his own project, as is the designer's, it is very important that a complete and clear understanding of his requirements be achieved.   (If the architect or engineer has done all of his homework he will probably know more about the job than the potential client does—but the humble, inquisitive approach helps to put the client at ease, bridge communication gaps, and should make him more responsive to the later formal presentation.)   Get the client to discuss his project, his problems, and what he thinks you might be able to do for him.   Have a few questions in mind to get this discussion underway; in effect, questions he should be asking himself.

If the potential client has a program—even partial—or preliminary sketches of what he thinks he wants, discuss these points
thoroughly. Tack his sketches, maps, or whatever he may bring
to the interview on the wall of the conference room, then encourage
him to make a brief presentation of his own. Have whatever
graphics have been accumulated by the design firm to illustrate
his site or project—even a topo map—mounted on the wall before
the client arrives. The professional is thus building bridges to the
client and encouraging a relation of the project to him as the
architect or engineer long before getting into the formal presentation covering the firm's experience, staff, and projects closely related to the client's.

Gradually lead the discussion around to your firm and your
experience with similar jobs. Mention a few past and current
clients he might know. Constantly emphasize and restate an interest in his project.

When the timing seems right, move into the formal presentation
of slides, motion pictures, wall displays, models, flip charts, large
mounted photographs, or whatever combination of these has been
assembled for the interview. If the visual presentation is primarily
a slide show, try to limit it to the firm's functional philosophy
(the "big idea"); understate "design." And keep it brief. There
is nothing quite so disconcerting as seeing the president of a
large corporation nodding in his chair in the middle of a slide
presentation. If the slide show appears to be running too long
or if the client seems to be losing interest, gracefully bring it
to a close.

When making a presentation on the potential client's home
grounds—particularly if he has not yet visited the designer's office
—it is not a bad idea to begin the slide show with a few views of
the professional's offices. Include photographs of the lobby, principals' offices, drafting rooms (with people at work), model-build-
ing shop, and any other unusual activities which can be explained
in one or two slides. A dozen such slides, at most, moved through
fairly quickly, will serve as a substitute for an actual visit, if necessary, or as a teaser for the potential client you want to get into
your office later on. Do not forget to show a few pretty women
staff members in the photographic tour—sex sells more than men's
magazines and black lace underwear.

SINGLE-PROJECT PRESENTATION

It sometimes is better to concentrate on covering one related job thoroughly with slides, rather than presenting a mélange of many different projects and trying to tell the story of each one with one or two slides. This single-project approach is particularly effective if begun with views of a master plan study, progressing on through concept studies, schematics, etc., to slides showing the finished and occupied building.

Do not take telephone calls in the conference room during an interview. It is all right for a secretary to come in once or twice during the interview with a note. Excuse yourself, read the note, then tell the secretary you will return the call later. Use your own judgment about explaining the caller to the client—embroider a bit on the situation and the importance of the caller if you like. By the same token, make certain that any outside calls for the visitors are relayed promptly to them.

This may be a bit reminiscent of the stories told by some of the architects who entered practice in the dark days of 1929 or the early thirties. There was little work and less money in that period and several of today's prominent New York City practitioners used to while away the long mornings and longer afternoons in a west side speakeasy. Occasionally, one of the group would get a potential client interested in coming in to his office for a presentation and tour of his quarters. An S.O.S. would immediately go out to the others in the speakeasy. Abandoning their drinks and the free lunch, they rushed to the lucky one's office and took up positions at the otherwise empty drafting boards. The potential client, knowing full well the state of the economy, had to be impressed with this evidence of his architect's busy practice. The next day, or the day after, the successful "principal" could be found manning a board for another of the group who had managed to find a live one.

GIMMICKS

Interview gimmicks abound and there is no reason to attempt to catalog them here; one example should serve to explain what is meant by an "interview gimmick." Name performers from night-clubs, television, the legitimate theater, summer stock, or motion picture units which might be shooting on location nearby often will make themselves available for a fee to "accidently" walk in on a presentation in progress. They will look properly startled, greet you by your first name or nickname, excuse themselves for barging in and mumble something about their "project"—leaving the impression that the two of you are working together on an unannounced multimillion dollar job. Your role is to assure him or her that the interruption was OK, introduce the performer to the client, and assure him or her that you will be in touch later about the shopping center or hotel or house or dinner—whatever seems appropriate. Personalities may be booked for such walk-on roles through their agents or talent agency; they pick up some easy money and you make an unforgettable (you hope) impression on a potential client. Fees for the talent could run from a few hundred dollars to several thousand dollars, depending on the star's popularity of the moment and how big a part he or she is expected to play.

EVERYONE PARTICIPATES

To return to the more professional aspects of the interview—be certain that each member of the staff present at the interview makes a contribution to the presentation. The presentation format will already have been established in murder sessions or similar planning meetings, long before the client appears. It is in these pre-presentation meetings that the part to be played by each participant is outlined. Likewise, attempt to draw into the discussion all of the client's representatives present at the interview.

There is a danger, particularly in presentations made in the client's office, in having too large a group representing the design firm. One prominent architect usually shows up for interviews with a retinue of eight to ten staff members. Most of the escort group seems to be there as highly paid porters, in that after the vast array of equipment—slide and motion picture projectors, screens, flip charts, models, brochures, and booklets—is set up, they retire to the sidelines and sit quietly while the principal makes the presentation. This overdisplay of manpower has cost the firm some jobs.

If the presentation team consists of two or three professionals (a good number), each man should feel completely comfortable in working with the remainder of the group.   Murder sessions or practice presentations should have helped in this regard.   Each presenter must be able to sublimate his individuality to some extent, feeding openings and cues to the others.   The remaining team members must then be able to recognize and pick up the lead-ins and move the presentation along smoothly.   This technique has been likened to the approach of an experienced police interrogation team, where one officer comes through to the prisoner as the "bad" guy, while his partner assumes the role of the "good" guy, attempting to establish a friendly relationship in the face of his partner's evident hostility.   After so long a time, such role playing becomes second nature to most police officers — and the same thing should be true of a presentation team of professional designers.   In the latter case, everyone should come through as good guys, of course.

Be well prepared for certain tough questions from the potential client.   (Review the questions for Doe, Roe & Smith in the preceding chapter.)

As another general rule, most firms do not include consultants in the presentation team.   Other offices bring them along as full participants on the basis that a consultant can be very helpful in selling a complex job.   This would be especially true when the consultant is a recognized authority in a particular project type.   If the election is to exclude consultants — at least in the initial interview — then be prepared with names of experts who would be used, in case the client asks.

COMMITTING STAFF

Clients often want to meet the people from the office who will be assigned as partner-in-charge or project architect on their job if the firm is selected.   Sometimes this will be made clear in advance of the interview, but the presentation team should always be prepared for the question by having some names at least tentatively in mind.   Be sure that any potential assignees are briefed on the client and his project.   Make a realistic selection from among principals and top staff members, based on current work flow, progress of jobs to which they are currently assigned, and their other in-house responsibilities. Above all, leave leeway in any commitment. Explain that you can be specific as of this week or next, but it is understandably difficult to commit men on speculation for six months from today. If the client presses for a concrete answer, give up any further attempts at evasiveness and tell him what he apparently wants to hear. These matters have a way of working themselves out.

TELL HIM WHAT YOU TOLD HIM

The summarization of a presentation is extremely important. The potential client must be reminded of all the important points he has heard in the past thirty to sixty minutes. Some salesmen refer to this as "telling him what you told him." An article with that title in *Marketing Times* made several worthwhile points (slightly adapted):

> When you make a presentation, everything you do leads, hopefully, to getting the [design commission]. You sell benefits, selling hardest on those most appealing to this particular prospect.
>   If your [presentation] is any good at all, it has proved those benefits, one at a time, step by step.
>   You've told the [potential client] everything to make him buy. Now, if you need to, tell him again.
>   Only this time, do it in summary form. If he's listened to your [presentation], you don't need the explanations used the first time around.
>   Nail the lid on the [presentation]—bang, bang, bang! . . . Use shotgun shells in your [presentation]. Use a machine gun in your summary.[4]

A sales manager is quoted on summaries in the same article:

> The sharpest summary ever devised won't sell a dime's worth of business by itself. Revise it every time to fit squarely into a selling situation. Start with a readymade summary, but make alterations to get a job. After I've gone through the benefits [for one client], I can always tell the one he's bought. When it comes time for the summary, I summarize that one benefit and forget everything else. He appreciates that I don't waste time on anything except his major interest. For another customer, I summarize every benefit for him. If I didn't, this fellow would feel shortchanged. He wants a mental file of all of the reasons he should buy.[5]

SELL WHAT YOU HAVE

To state the obvious—in every presentation a firm must sell what
it has to sell, and sell against its known competition.   Part of the
intelligence-gathering process is learning the names of potential
competitors.   If possible, get from a representative of the potential
client his original list of firms before any cuts are made.   This
will give a firm at least two advantages during the selection process: (1) it will make it much easier to keep track of the firms
eliminated in each cut by the client and (2) the composition of the
starting lineup may contain important clues to the type of design
firm the client believes he wants for his project.

As for selling what you have, there is an infinite number of possible combinations of in-house disciplines, experience, and other
selling points with which a client may be confronted in interviewing six to ten design firms.   Some offices have no engineering
capability in-house; others have practically every engineering
discipline known to man under one roof.   The first office will
carefully explain to the client the advantages of using engineering
consultants, while the better-endowed firm will make a very con-
vincing case for having all engineers at hand on a full-time basis.
The locally based firm must convince the client that it can best
serve his requirements by being available around the clock; the
out-of-city or out-of-state designer makes the case for jet airplane
mobility and other points for *not* being on the scene—all in the
best interests of the client.   Firms with a long list of design awards
naturally dangle their medals and certificates in the client's face;
the award-shy office points out the dangers to client budgets of
being too "design oriented" (whatever that term may really mean).
If one has several scholarship winners on his staff, this point is
subtly made during the presentation—and so it goes.

Although practically all of the arguments and counterarguments
for these and many other points are known to the author and many
readers of this book, there will be no attempt to list them here.
On reflection, it might be interesting to compile all of them at some
future date, if it were not for the total confusion such a list could
inflict upon potential clients who might happen to see it.

Sit in on presentations by other offices whenever the opportunity
presents itself.   Even though such chances rarely come to most

practitioners, this is perhaps the best way to rate your own presentations.

FOLLOW-UP

Immediately following the interview, write the client to summarize in permanent form all the major points which were covered in the interview. Sometimes the follow-up letter takes the form of or incorporates a proposal letter. Client desires, as expressed during the interview, usually will govern the format. Always restate your interest in serving the project. Include any additional reprints or other supportive material with the letter which you believe could strengthen your case. One firm illustrates its close attention to detail by including a copy of "Building Program Considerations," developed several years ago for a specific client and now refined into a general promotion piece. The first page of "Building Program Considerations" looks like this:

BUILDING PROGRAM CONSIDERATIONS

1. Generally, what are the facilities, operations and functions to be housed? What is the general scope by area and function (existing and projected?)
2. Are there any specific factors which will control or dictate the sequence of the construction; functional interrelationships by department or operation or similar considerations?
3. Has a construction budget been established? If so, to what extent has this been specifically coordinated with requirements?
4. Has any specific design direction been established or indicated by policy decision, unique requirements, etc.?
5. What is the client's organization for the departments or operations involved in this project?
6. Provide a list of all personnel with whom the designer will work, including titles and responsibilities.
7. How much of the following data is already available?
   a. Employee count by department. Include equipment count where applicable.
   b. Area requirements by department based on employee count and individual space requirements.
   c. Functional adjacency requirements by departments and/or process —primary, secondary, etc.
8. To what degree do employees circulate between departments or other organizational units within the project and/or the existing building?

9. What are the requirements for and how extensive are the shipping and receiving facilities contemplated? Service pickup and delivery? What size and weight of trucking is expected?

This should be enough to give the general idea of the approach.

For those readers who might be interested, here is a sample letter to a potential client following a presentation:

Mr. Arthur R. Stein
214 Saxon Avenue
Kittsford, New York

Dear Mr. Stein:

On behalf of George Meltz, Carson Jones, and myself, I want to thank you for the opportunity to meet with you and the other members of the Building Committee on June 00. I trust we made clear during our discussions that we are very interested in serving as the architect for your project. I am confident that we can provide the necessary experience and range of services required by your congregation, and we believe we are in a unique position to respect the important historic and esthetic design considerations which arose during our meeting.

To recap briefly a few of the major points which were raised during the meeting, for your consideration in reaching a decision about an architect:

1. Fiscal responsibility. It is no longer fashionable—nor has it been for many years—for an architect to approach the solution of his client's design problems without paying close attention to meeting the budget established with the client. We are not interested in fashioning beautiful plans which do not get built. Our fiscal responsibility to the client is based on an in-house estimating department of experienced men, most of whom we have brought in from the construction industry. Cost estimates are checked and rechecked all through the design and construction phases of the job.

2. Humanity in design. An unfortunate but common error is to forget that buildings are not commissioned by clients to be monuments to some architectural office—and, even worse, to ignore the fact that people will use the building. Achieving humanness in scale, in function, in esthetics generally, is the continuing assignment and challenge to today's responsible architect. We do not have standard styles or pat answers to design problems; your project is designed with the considerations characteristic of your program, your requirements and your site and neighborhood always in mind.

3. In the case of your proposed new structure, we are very conscious of the need to protect the residential character and outstanding landscaping of your site.

4. The first obvious step for your committee is to consider a master

plan study of your total needs. This can be accomplished in stages if financial or other considerations so dictate. In this connection, I would like to re-emphasize the point that we provide a full range of services complementing our design capabilities. These important design-related services include site engineering and landscaping, interior design, estimating, mechanical and electrical engineering, computer analysis, and field supervision.

From our discussions I estimate we will need four to six months to develop the programming and schematic design. This would include production of a concept model and other visual aids required for your fund-raising efforts and for presentation to local authorities. Under Carson Jones's direction we have a fully staffed and experienced communications department to work with your fund-raising company as necessary or desired.

Our fees are commensurate with the recommended minimum fee schedule of the New York Society of Architects, dependent upon the scope of total services we perform.

As I promised you, I have enclosed some of the program requirements and questionnaires prepared in connection with our studies for my own congregation, Old York Road Temple in Abshire, Illinois. As discussed during our meeting, I would be available personally to assist with your planning, so as to give you the benefit of my experience on a building committee similar to your own.

Should you or any members of the building committee have any questions which we can answer at this point, please let me know. When you are ready to visit some of our projects and talk to past clients, George, Carson, or I would be glad to make the arrangements. We would also like to re-extend our invitation to visit our offices in Chicago.

Sincerely yours,
ARTHUR K. SMITH & ASSOCIATES
F. T. Ironsides, AIA
Associate

After sending the interview follow-up letter, with whatever enclosures seem appropriate, ease off on all external pressures on the potential client. Again, these are *general* rules; a continuation or even an intensification of personal, business, or political contacts might be indicated in some cases. If an office finds that it cannot serve the project for any reason, the principals should immediately withdraw the firm from further consideration in a diplomatic and professional manner.

When the job goes to someone else, which will be 75 to 90 percent of the time, try to find out why—particularly if several cuts were survived before the formal interview. Unbelievable as it sounds, many offices drop all further contact with potential clients as soon as they learn that they were passed over for a particular project. It is only simple courtesy to thank the client for his consideration and visit, if there was an interview, and it is good basic public relations for future consideration by the client and his associates. If it can be learned from the client or from other sources which firm was selected, assure the client that he made an excellent choice, preferably in writing with a copy to the principal of the firm that was selected. Such a gesture costs nothing, is a sign of professionalism, shows that you are a good sport to the prospective client—for whatever that could be worth when another job comes along in his organization—and is bound to leave a lasting, favorable impression on the winning designer.

Finally, maintain contact with all clients and potential clients, regardless of the outcome of any one presentation. If the approaches are kept open to the bridges which were painstakingly built over the weeks and months leading up to a presentation, the possibilities for consideration for future projects should be excellent.

REFERENCES

[1] Warren Leslie, *Under the Skin*, Bernard Geis Associates, Inc., New York, 1970, pp. 150–151.
[2] George T. Vardaman, *Effective Communication of Ideas*, Van Nostrand Reinhold Co., New York, 1970, p. 89.
[3] Ibid., pp. 131–132.
[4] George B. Anderson, "Tell Him What You Told Him," *Marketing Times*, New York, January–February 1972, pp. 10–14.
[5] Ibid., pp. 13–14.

CHAPTER 9
Political Action

Like "propaganda," the words "lobby," "lobbying," and "lobbyist" have unfortunately gained rather shabby connotations among English-speaking peoples. Many persons automatically relate lobbying to the activities of five-percenters and other types of anti-social freebooters who prey on society and its institutions.

In a government affairs pamphlet the AIA has defined lobbying as "the attempt to influence legislation advantageous to a particular interest. It is the supplying of detailed professional information — or opinion — to legislators and their staff."

Webster's Third New International Dictionary says lobbyists are "the persons who frequent the lobbies of a legislative house to do business with the members; specifically, persons not members of a legislative body and not holding government office who attempt to influence legislators or other public officials through personal contact."[1]

Still another definition — this from a practicing lobbyist — has it as "trying to influence the passage of national or state legislation, positively, so that it will be beneficial to, or negatively, not detrimental to, the interests of the client or clients who pay the fees for such services."

Professional lobbyists, incidentally, usually prefer to be known as legislative representatives.

LOBBYING ACTIVITIES

This particular legislative representative recently explained his activities over one thirteen-month period, all directed at getting one word deleted from legislation pending before congressional committees. The word, naturally, was a very important one and had it remained in the bill, would have had an extremely adverse effect on his client's business. His actions on behalf of his client in this case included:

1. Writing to the members of the appropriate congressional committees—two in the House and one in the Senate—and providing them with an intelligent analysis of why the word should be deleted from the bill under consideration. This activity was repeated several times, as the climate changed and necessity dictated.

2. Appearances at two committee open hearings as an informed (expert) witness, and repeating the story of why the word should be dropped from the bill.

3. Corresponding with committee counsel and committee chairmen regarding the general ramifications and impact of the proposed legislation.

4. Preparing and distributing fact sheets and position papers to key people in the industry suggesting amendments to the legislation, as opposed to attempting to block it entirely.

5. Writing occasional key articles and making speeches to industry groups.

The result of these activities was a bill without the offending word, subsequently enacted into law. The lobbyist gives a few additional pointers about influencing legislation and legislative minds:

HOW TO LOBBY

1. When not too pressed by the demands of constituents and the need for re-election, most state and national legislators have open minds and try to do their job well.

2. They will welcome and consider carefully any suggestions about pending legislation which are not obviously destructive, supremely selfish, or demonstrably ignorant of public need.

3. They are human and don't want to do anything or take a stand which might later be considered to be foolish, shortsighted, or ignorant of true conditions.

4. They are more impressed by and attach more credibility to facts and figures from parties affected directly by legislation than from lobbyists who work for hire.

5. Generally, legislators are open-minded on amendments, but "stop-legislation" movements tend to arouse their ire, particularly if the legislation is popular with the public. "Stop-action" lobbying also sometimes carries with it an implication that legislators who favor the legislation are stupid or venal or both—impressions which tend to make the lobbyist unpopular.

Further advice to lobbyists—professional and amateur—is contained in *Ten Commandments of an Effective Lobbyist,* distributed at the AIA-CEC Public Affairs Conference in Washington, D.C., in March 1972:

1. BE HONEST. Be straightforward in presenting your view, give both sides of the issue, do not imply facts or authority which you do not possess.

2. BE INTERESTED. Know your legislator, familiarize yourself with his problems and specialized interests, meet his staff, stop in, and maintain frequent and friendly contact. (Example: Ask to be placed on his constituent newsletter list.)

3. BE CONSTRUCTIVE. Don't just condemn a proposal, show specifically how it can be improved, demonstrate a willingness to compromise if compromise can be found and take positive, as well as negative, stands on issues.

4. BE ALERT AND IMAGINATIVE. Sell your position or proposal through surveys or programs which catch the public eye. If you favor a measure seek a means of dramatically promoting public interest and support. Watch for opportunities to pursue your proposal in other ways, perhaps as an amendment or resolution, and watch for amendments or resolutions which affect you.

5. BE COOPERATIVE. Coordinate your legislative activities with other groups or organizations which may share your interest in a given bill or proposal. This includes organizations which might be expected to oppose, as well as support, your position. Also, keep abreast of legislation in other states, plus national legislation studies affecting your industry.

6. PROVIDE INFORMATION. Give your legislator fact sheets, background studies, position papers and similar research which tells how a proposal affects *you.* Then provide information on how it affects the public and/or your industry and *keep it brief!* If the legislator wants more, he'll ask for it.

7. PRACTICE WHAT YOU PREACH. Your words are worthless

if your acts contradict them. (Example: Don't say A-E's can't bid, and then submit a priced proposal.) Exemplify the position in all that you undertake.

8. BE PROFESSIONAL. Don't threaten your legislator if he disagrees with you. By the same token, never hint at rewards for his support. Never get angry, and *never* "try to get even."

9. AVOID STANDARDIZATION. Do not send mimeographed or dittoed letters. Do not use form letters provided by your associates. Never provide copies of letters from your association asking you to act. Send original letters in your own words.

10. REMEMBER TO SAY "THANK YOU." After an issue has been resolved, send a note of appreciation to your legislator acknowledging his time and effort in your behalf. Do this regardless of extent of such effort. Give the legislator the benefit of any doubt.

WHO SHOULD LOBBY?

The New York Times[2] for October 18, 1971, carried an interesting case history illustrative of the lobbyist's point about the relatively greater impact on legislators of parties affected by proposed legislation, rather than hired advocates.

A member of the House Ways and Means Committee received a letter from the head of a small textile plant in his district, asking for certain exemptions in the proposed investment tax credit legislation on domestically produced machinery. "It was the simplest, most common and, in many instances, the most effective form of lobbying," the *Times* story pointed out, "a letter from a constituent who wanted help to a knowledgeable and influential Congressman who could provide it."

> In the case of (the) Representative . . . , the brief letter from his constituent was all that he needed to understand the situation. Textile companies are the largest employers in his district, and, in 18 years in Congress he has become expert in their problems.

The same article explained why America's regulated utilities were represented before the Ways and Means and Finance Committees of the House by the president of a tiny independent telephone company in Arkansas. "It would have been counterproductive for AT&T to lobby directly," a utility industry source explained. "It's an enormous company and nobody bleeds for the big company. But the little fellow from Arkansas—a member of Congress can relate to this guy and might want to help him."

For a detailed insight into the world of lobbying and influence peddling, with names, dates, and details, read *The Washington Pay-Off*—a 1972 lid-blower by former lobbyist Robert N. Winter-Berger.

By this point you may be asking yourself just what does lobbying have to do with getting new work—"selling professional services." It is doubtful that many aware members of the design professions would deny that governments at all levels are the largest single client body today, with every sign that this situation will become even more true in the last quarter of the twentieth century. The next chapter, "Federal, State and Municipal Clients," should win over any remaining nonbelievers.

PROFESSIONAL LOBBYING

At the 1971 AIA convention in Detroit, as those readers who attended may recall, President Robert Hastings urged "architects and the Institute itself to enter the political arena, enlist allies, swing votes, mobilize community action and take positions on issues heretofore considered outside the purview of the design profession."

The AIA president was calling for action or a commitment on the part of the design profession to, among other things, preserve and restore our decaying environment—both natural and man-made—and to alter "a wide range of public institutions that are failing to respond to demonstrable public needs."

All very well—and while this is the sort of rhetoric one expects to hear in conventions of architects, engineers, landscape designers, and others in the design professions, such generalities are not what this chapter is about. What it is about is personal, individual political action and activities by the professional, directed at making lasting and productive contacts and obtaining guidance, assistance, *and jobs* in city halls, county courthouses, state capitals and in Washington, D.C. In short, there will be found herein no lofty explanations of the architect's role in government or social reformation; no inspired exhortations to mount up and go atilting at legislative windmills on behalf of the oppressed and down-trodden of the world. In practical politics, as in business development, one deals with what exists here and now.

POLITICAL INVOLVEMENT

"Politics is everybody's business" is a trite but terribly true axiom for our times. Those who have not only heard the admonition but also heeded and acted upon it, are mostly in better shape today, practicewise, than those who were too busy or just not interested in dipping more than the end of their big toe into politics. All that was said earlier in this chapter about lobbying has a direct relationship to individual political action on one's own behalf. Personal and political relations are difficult to divorce from each other; when you talk to your congressman about your interest in a new federal building or a post office renovation in your hometown you are inevitably engaging in a form of lobbying.

A currently popular "in" word is "involvement." Individual involvement is exactly what is required to understand and benefit from the political process in America. Unfortunately, involvement is also the most missing ingredient.

WORK FOR YOUR CANDIDATE

Elected representatives—city, county, state, and national—always appreciate an individual's vote and those of as many others as he can influence at election time. A much more welcome (and provable) form of support is active participation in the campaign, beginning with the primaries, if one really wants to stand up and be counted. If you don't have the time or temperament to become a volunteer campaign worker, the next most appreciated kind of involvement is a contribution of money. Liken a primary win to the importance of surviving the client's last cut and being scheduled for an interview. Then compare winning in the general election to being awarded the design commission.

Talk is no longer cheap. This is nowhere so true as in modern political campaigns, where even a locally broadcast sixty-second commercial on television can cost a candidate thousands of dollars for air time and production. Campaigning requires vast inputs of volunteer hours and money; winning demands even more of both—and the name of the game in politics is winning.

In a May 1972 article in *TV Guide* it was pointed out that "in 1968 a total of $300 million was spent on political campaigns, a 50-percent increase over 1964. Of that amount, broadcasting expenditures accounted for $58.9 million, an incredible 70-percent jump from 1964's $34.6 million. In the presidential race alone, about $12.6 million was spent for broadcasting by the Nixon forces, and about $6.1 million was disbursed on behalf of Hubert Humphrey."[3]

Not being able to find a candidate you can wholeheartedly support is no excuse for nonparticipation. Every fourth year, in presidential campaigns, there are more than 8,000 major political offices to be filled at state and national levels. In 1972, for example, there were 7,326 state legislative races (1,902 Senate, 5,424 House), 20 gubernatorial races, 35 U.S. Senate seats open (14 Democratic and 19 Republican incumbents), 435 House of Representative seats and, of course, the Presidency and Vice Presidency of the United States. This does not take into account the thousands of other city, county, and state elective positions which are also voted on in presidential election years.

In off-year elections a proportionately large group of candidates likewise seeks elective offices.

If, among this vast group of political races and candidates, at least one cannot be found for whom you can sincerely and unreservedly campaign, then there is the option of generally supporting one of the parties by working or contributing funds or both.

BE YOUR OWN CANDIDATE

Up to this point the discussion has centered around moral and financial support of candidates by design professionals. As must be evident, any adult citizen of the United States who has not been deprived of his franchise through being convicted of certain serious crimes is free to toss a hat into any political ring which appeals to him. There are age minimums for some legislative and executive posts, but these should be no bar to most architects, engineers, or planners who have been in practice for a few years.

Some of the professional and trade publications examined this subject in 1971 and 1972. *Progressive Architecture* was one of the first publications to carry a substantive story about design professionals in elective offices; *Building Design and Construction* took up the subject in its April 1972 issue.

A governor's mansion was about as high as anyone had reached by 1972. Tom Judge, the former executive secretary of the AIA's

Montana chapter and a nonarchitect, moved up from lieutenant governor to Governor of Montana. Several architects and engineers have made it into their city councils and state legislatures. No designers have yet made it to the U.S. Senate or House of Representatives, but several have been candidates in recent years. According to *Building Design & Construction,* two of the engineers who ran for congress in 1970 lost both the election and their practices due to the financial inroads of campaigning.

Unfortunately, the design profession has not yet seen fit to support their own in political campaigns, unlike the medical and legal professions. The director of the Consulting Engineers Council was quoted as saying, "One of the engineering candidates said he got more campaign contributions from the beauty shop operators in his state than from consulting engineers throughout the U.S."

The Political Committee for Design Professionals was organized in 1972 to "raise funds to support candidates running for the United States Senate and House of Representatives who understand and support the goals of architects, engineers, planners and landscape architects." The eight PCDP trustees—three architects, three engineers, one planner, and one landscape architect—solicit funds and determine the final distribution of money to political candidates in both primary and general elections.

The 1972 fund goal was $250,000, but since the first formal solicitation letter did not go out until mid-October, it is doubtful that the goal was reached. Contributions of over ninety-nine dollars were subtly discouraged to avoid public identification of individual donors. At least it is a start by the design professions toward building up a campaign kitty to support candidates friendly to the goals of the profession. It is to be hoped that the trustees will find at least some of their own members who are deserving of financial backing in future campaigns.

Some AIA leaders have had an interesting outlook on architects running for public office. *Building Design & Construction* quoted a former president: "Although the AIA is happy to see its members pursue elective public positions, most AIA officials think that greater change can be produced by broadscale citizen involvement. By remaining within their expertise, A-E's can deal from a position of strength and retain credibility."[4] Such a head-in-the-sand attitude on the part of their professional organization could be even worse for architects than the once-fashionable ivory-tower outlook.

ON CAMPAIGNING

A professional campaign worker made these pertinent comments and observations in a 1971 political workshop:

> A lot of people, man-hours, money and plain hard work go into political campaigns in our country. Yet even to many thinking people the process whereby a political candidate campaigns and either wins or loses his bid for public office is an enigma.
>
> It has been demonstrated time and again that an essential ingredient of winning elections is a well planned, well organized campaign that makes full and complete use of every source of potential strength. At all times keep in mind that the only purpose for projects, ideas, promotions and scheduling is a means or methods for achieving victory. Each project and each dollar must be measured against one basic standard or guideline—will it produce votes? A political campaign is not designed to "get across a certain issue" or "educate the voters"—it is to elect your candidate to office. A loser never educates or legislates. By organization, planning and follow-up you can get your workers talking enthusiastically and favorably about the candidate and the campaign. Having people "talk in friendly terms" is the most effective winning campaign technique devised. Always impress on every worker: *The sole aim of the campaign is to win.*[5]

If you opt to work for a particular candidate remember that each candidate is different; each has his strong points and weaknesses. Each has broader knowledge in some areas than in others; each has particular personal or intellectual attributes which can be used to good advantage in a campaign. But all candidates require a sound, competent campaign organization.

CAMPAIGN PLANNING

The first task of the campaign committee is to work out an overall campaign plan. In the first campaign committee of which I was a member—some sixteen years ago—assembling a campaign plan of thirty-two pages plus ten pages of appendixes was the first step. The plan's divisions covered:

1. The Incumbent
2. The Candidate
3. The Campaign
   a. The Primary
   b. The Facts
   c. The Consequences
   d. Plan of Attack
   e. The Candidate's Personality

   f. The Positive Side
   g. Caveat
   h. Pacing
   i. Appeal
   j. The Party
  4. The Opposition's Campaign
   a. Smears
  5. Conclusions

The appendixes included several versions of fund-raising letters with follow-up reminders, together with a recommended format for campaign press releases.

The "Facts" section of the campaign plan contained an estimate that our man would need 95,000 votes to win; his final total was 95,346 votes. (He won.) The campaign plan, along with most of the correspondence from the campaign committee, is considered highly confidential. It is good practice to control all copies of the plan, for instance, by numbering each copy and recording the name of the committeeman to whom it is issued.

This particular campaign example serves to illustrate several of the points made so far in this chapter. The candidate, who had never previously stood for elective office, was running for a seat in the U.S. House of Representatives. He was subsequently reelected to Congress several times and during a distinguished political career served as a presidential assistant in the White House. Later he was appointed a U.S. ambassador to a prestigious post in Europe. Obviously, one cannot predict that all of the candidates for whom one works will enjoy such a meteoric rise in government—but the possibilities are always there.

PARTY ORGANIZATION

On the other hand, if the decision is to work for a party rather than a specific candidate, one should be familiar with the basics of party organization. Both major parties are organized pretty much along the same lines. They are made up of committees: precinct, county, state, and national. These committees are governed by state law or by rules established in the respective national conventions. Some states organize by wards, election districts, towns, and cities, but for our purposes the precinct-county-state-national designations will be used.

The precinct committee is headed by a committeeman and/or woman elected by party members of the precinct at a precinct meeting, in a primary or regular election, or as otherwise provided by the party's state rules. The precinct committee is responsible for seeing that all of its party's voters are registered and vote on election day. It is the unit of political organization closest to the voter.

The county committee is elected by the precinct committeemen and women, by the voters of the party in the county, or as otherwise provided by state law. It consists of committeemen and committeewomen and other necessary officers. The county committee must see that candidates are selected for local offices and is responsible for the campaign of these candidates as well as for the vote in the entire county. The county committee must also see that every precinct has active leadership, that the leaders are supplied with information and instructions, and that on election day every polling place is covered by well-trained election officials.

The state committee is composed of members—usually a man and a woman—representing the counties or other political subdivisions in the state. The state chairman and vice chairman are chosen by this committee, which has the responsibility for conducting statewide campaigns and for building the county organizations.

The national committee is made up of a man and a woman from each state and territory who act as the governing body of the party on the national level. This is the committee responsible for organizing the national convention, directing the presidential campaign, and formulating national campaign programs to be carried out in the states. The national committee also is charged with encouraging and assisting the states in building their organizations.

The national convention, made up of delegates from all states and territories elected by state primary, state convention, or state committee, nominates candidates for president and vice president, adopts a party platform, and is the ruling and governing body of the party.

With that outline of how the national parties organize and function, let's turn to the usually taboo subject of political contributions.

CAMPAIGN CONTRIBUTIONS

There are laws at most levels of government which attempt to govern the amount of campaign contributions made and what the recipients do with the funds. The Federal Corrupt Practices Act of 1925, as amended from time to time, set out specific rules on the maximum amount of contributions, reporting, and other matters pertinent to campaign collections and spending. In the past, such limitations were often circumvented by setting up a multitude of supposedly independent committees—especially in the District of Columbia—which then received and spent the maximum of $3 million each, but all for the same candidate. A wealthy contributor could give the $5,000 individual maximum to any or all of the committees working for his candidate. As Tom Wicker once wrote in his *New York Times* column, "Everyone who engages in, writes about or has anything to do with politics knows that these statutes are as violated as the Ten Commandments, and far more systematically."[6]

On April 7, 1972, the rules of campaign contributions underwent drastic changes, with the implementation of the Federal Election Campaign Act. This legislation was the first overall reform of the system attempted since the above-mentioned Federal Corrupt Practices Act. As of the April 1972 effective date of the new law, contributions of more than $100 to any candidate had to be reported by name, address, occupation, and place of business of the contributor. Any committee which expected to receive at least $1,000 for a candidate must furnish the same breakdown of all contributors. The mandatory reports are due several times during the year and in election years reports are required fifteen days and again five days before each state primary race.

Washington lobbyists immediately began an intensive search for loopholes in the new law. There seem to be few of any substance, but as one veteran of political campaigning put it: "The same old loophole—cold, green cash—is still with us."

An understandable summary of the new Federal Election Campaign Act is carried in an unusual book called *The Political Marketplace*. Edited by David L. Rosenbloom and published at $25 by Quadrangle Books, New York City, the 1972 edition combines explanations of many aspects of the political process with a directory of campaign services and lists of party and elected officials from all over the United States. Contents include:

6,000 county chairmen
7,000 state legislators
  500 statewide elected officials
  500 mayors of major cities
  300 state and national political committees
  300 campaign management and supply firms
2,000 major media outlets
Opinion polling and voter research firms
Film producers
Fund raisers
Telephone consultants
Bibliography of courses and books on campaigns

Editor Rosenbloom explains the reason for a book like *The Political Marketplace* in his introduction:

> The campaign planning indexes in *The Political Marketplace* indicate fully the kinds of services now being offered to candidates by professional management firms. The planning indexes are in fact a checklist of the things that must be done in modern campaigns. Some of the firms do all of these things, others specialize in one or two. Even if a campaigner decides to run his own campaign with friends and party workers, he will find *The Political Marketplace* indexes to be a useful planning aid. They provide a map of the things he will have to do to wage an effective modern campaign.
>
> Publication of *The Political Marketplace* shows very clearly that politics is different today. That does not necessarily mean that it is better or more democratic. In fact, many scholars have described the changes marked by this publication as the disintegration of American electoral politics and parties. They argue that commercialized politics gives us the form but not the substance of real political contest.
>
> It is too soon to say whether they are right or wrong. However, if they are correct, our effort to bring all the elements of *The Political Marketplace* together in one source volume will have commemorated the collapse of American democratic politics as we have known it. Politics may have become a big business and not much else.[7]

How much, if anything, you as a design professional contribute to political campaigns or candidates should be a matter for individual decision and conscience. Most of the professional societies provide guidance on the subject in one way or another in their ethical practice standards.

The National Society of Professional Engineers' Code of Ethics, in Section 11(b), states "[The engineer] will not offer to pay, either directly or indirectly, any commission, political contribution, or a gift, or other consideration in order to secure work, exclusive of securing salaried positions through employment agencies."

It is perhaps indicative of the greater importance of political action to today's designer that the AIA's original Canons of Ethics (Chapter 1) was silent about contributions, political and otherwise, in 1909. Nor did architect-author Royal Barry Wills, cited in earlier chapters, have anything to say on the subject as late as 1941 when his book, *This Business of Architecture*, was published.

CONTRIBUTION SCANDALS

Political contributions by architects and others in the design fields were a subject of hot controversy in some areas in 1970–71. This was particularly true in New York State where a Long Island newspaper, *Newsday*, uncovered enough evidence of questionable practices by some architectural and engineering firms to warrant a county grand jury investigation.

In its own nine-month investigation, *Newsday* found that at least $833,330 was collected in the six-year period ending in 1969 by the Nassau Democratic Party from architects and engineers who benefited from more than $36 million in contracts awarded by the administration of Nassau County Executive Eugene Nickerson. (It might be wondered whether the architectural and engineering firms reportedly involved were not to be more pitied than censured. According to the newspaper's figures, the total contributions — $833,330 — represented more than 40 percent of the fees which could have been realized on $36 million worth of design contracts. Few design firms could operate for long with that division of fees. If, on the other hand, the $36 million represented gross fees rather than construction costs, then the "contribution" figure drops to about 2½ percent — a much more livable amount.)

CONTRIBUTION CODES AND GUIDELINES

Giorgio Cavaglieri, then president of the New York City 1,600-member AIA Chapter, wrote his membership in early 1971 about the chapter's newly adopted ethics rule to limit members' political contributions:

The system of political campaigning and campaign financing now prac-

ticed in this country appears to be inextricably connected with the development of different kinds of collection systems, such as in Nassau
County, directed at those who could benefit from election of specific
candidates. . . . The currency of this issue as a matter of public concern, and our Chapter's involvement as a consequence of the *Newsday*
series, presents us with the special opportunity to lead and perhaps
break new ground as citizens in the professional practice of architecture.

The New York chapter's campaign contributions code, which
it attempted to have adopted nationally at the AIA's 1971 Detroit
national convention, has four major points:

- Member firms are prohibited from making political contributions.
- Individual members and their families are requested to limit contributions made to all candidates of any one party in a state to an annual
total of $500.
- All members are directed to file an annual report of political contributions made by them and their household families. If less than the
$500 total has been contributed the reports need not be itemized.
- Publication of all such reports is authorized, as well as publication
of the names of architects who fail to file.

Likewise prodded by the *Newsday* disclosures, the New York
State Association of Architects, Inc. took the leadership in the
formation of an ad hoc Joint Ethical Practices Committee, which
included the following groups:

American Institute of Consulting Engineers
American Society of Civil Engineers
Consulting Engineers Council—New York State
Long Island Chapter/Consulting Engineers Council—New York State
New York Association of Consulting Engineers
New York State Association of Architects, AIA
New York State Society of Professional Engineers

A statement on political contributions by architects and engineers was one of the results of the deliberations of this prestigious
committee. Pointing out that adherence to ethical standards is
the hallmark of a true professional, the statement continued:

One of the tenets of American democracy is the right of an individual
to support the political candidate or party of his choice. When this is
done voluntarily and without expectation of special favor or privilege,
there is no question of ethics. A definite conflict of interest arises,
however, when an architect or engineer responds to solicitations for
political contributions related to professional work.
These guidelines are provided to assist architects and engineers who
are confronted with such solicitations:

1. An architect or engineer may, as an individual, make contributions of service or of anything of value to those endeavors which he deems worthy, but not for the purpose of securing a commission or influencing his engagement or employment. The New York State Commissioner of Education defines as unprofessional conduct: "Offering to pay, paying or accepting either directly or indirectly, any substantial gift, bribe, or other consideration to influence the award of professional work."

2. An architect or engineer who believes that he has been denied professional work because of his refusal to contribute to a political activity or entity shall have the professional responsibility to report such circumstances to his professional society. Upon verification, the professional society shall bring the matter to the attention of the proper authorities, and, if appropriate, to public attention.

3. An architect or engineer who makes a direct or indirect contribution in any form under circumstances that are related to his selection for professional work shall be (a) subject to disciplinary action by his professional society, and, if appropriate (b) reported to the public authorities.

Delegates to the Detroit AIA convention generally supported limitations on the amount that can be spent in political campaigning for national offices, but referred the New York chapter's resolution, based on its new ethics rule, back to the AIA Board.

Meeting in Minneapolis in late September 1971, the AIA Board of Directors passed a resolution on political activity which said in part: "The Institute specifically condemns and holds to be a violation of its ethical standards (number 8) any political contribution of money or kind on the part of any member if such contribution be predicated upon an award or promise of an award of commission." The Board stated further that ". . . the Institute strongly supports federal, state and local legislation which has as its objective limitation, control and disclosure of campaign expenses and political contributions."[8]

None of these exhortatory statements appears to say any more or any less, or to be more specific, than the AIA's current Standards of Ethical Practice: "An architect may [not] make contributions of service or anything of value . . . for the purpose of securing a commission or influencing his engagement or employment."

Nor, unfortunately, does the September 1971 AIA Board action give any guidance about an equally sticky situation: when a representative of a government client threatens to cancel a major commission, already awarded, if the architect fails to make a campaign contribution. This happened not long ago during a mayoralty race in one of our larger cities when the favored candidate's campaign manager learned that one architectural firm was supporting his opponent. The job at stake involved millions of dollars in design fees, with several years to run to completion. In this case the architect involved immediately saw the advisability of making an equal (and substantial) campaign contribution to the other candidate, who went on to win the race.

Part of the ethical considerations involved in this situation concern the high added costs which would have fallen on the city's taxpayers if the architect had been changed in midstream. In the final analysis, who is to say that factor was not what motivated the architect's matching campaign contribution?

ENFORCING CONTRIBUTION CODES

The real problem in this area is that of enforcement. The local chapters of most professional societies do not have the power to impose or enforce sanctions against their members. Any member of the New York City AIA Chapter who violates the new code would be subject to censure, suspension, or the loss of his Institute membership only through action of the AIA's national judicial board. Verification of violations, followed by notification of "the proper authorities, and, if appropriate, [bringing] to public attention," as called for in the New York State ad hoc committe on ethical practices, would seem to leave any professional society taking action open to much more liability than its members are ready to assume.

A wry aside was provided by a spokesman for the Long Island (Nassau County) AIA Chapter, none of whose members were involved in *Newsday's* revelations. Explaining that his chapter had been awaiting "action by chapters whose members were involved," he predicted approval of a political ethics code by his membership.

GARDEN STATE POLITICS

A few months after *Newsday* opened up Nassau County's can of political-professional worms, federal indictments were handed down against several New Jersey politicians involved in the construction of a $40 million U.S. Postal Service facility in Kearney, New Jersey. An architect was named as coconspirator but was not indicted.

The indictments charged certain Hudson County political figures with plotting to shake down the developers of the parcel post building in order to get the contract.   Convictions were obtained on federal extortion-conspiracy charges.

The Postal Service appears to encounter more than its share of problems in locating buildings in Hudson County in the Garden State. *Engineering News-Record,* in the issue for May 20, 1971, pointed out that the cost of the Jersey City automated bulk mail handling post office had escalated some 52 percent in less than two years. This was the first of twenty-one similar mail handling stations to be built throughout the country under a $1 billion program. Cost of the 1.46-million square foot building, described by postal officials as an industrial-type structure, exclusive of land and equipment, would thus be about $42 a square foot.

A House of Representatives post office subcommittee, concerned about the cost estimates for the Jersey City facility zooming up from $62 million to $130 million—a 109 percent cost overrun, according to Rep. H. R. Gross of Iowa—began hearings on the project in late 1971.   The subcommittee's figures indicated costs of about $89 a square foot.   Disappearing records and reluctant witnesses marked the conduct of the House hearings.   The General Accounting Office characterized the mailing plant in the Jersey Meadowlands as ". . . shot through with blunders that border on criminal negligence, nonfeasance, malfeasance, or all three."

Topping off the disclosures in the House committee hearings was the admission that highly toxic and explosive methane gas was found beneath the construction site after nearly eight acres of concrete flooring had been poured for the massive building.

In still another federal grand jury investigation of affairs in Atlantic City, New Jersey, twenty-seven construction, service, and supply firms from Atlantic City and Philadelphia were subpoenaed. A prominent Philadelphia architectural firm was one of those called for an appearance before the grand jury in Newark.

That there are occasional violations of the laws and ethical codes governing political contributions is neither debatable nor here condoned.   It is, of course, not the action of a responsible citizen to dismiss evasion and outright violation of our laws by saying "everyone does it" or "the only real sin is in getting caught."   Involvement with political parties at the grass-root level by a much larger segment of our citizenry, as urged earlier in this chapter, is part of the answer.

SUPPORTING YOUR MAN

"Make your political contributions count" is the advice of experienced politicians and seasoned campaign contributors.   Decide on a campaign contribution plan in advance, with something put aside for contingencies.   Then stick to it.   Once you have decided to support a party and/or a candidate in a fairly substantial way, don't make the mistake of relaying your check through some underling or outlying committee member.   Request—insist—that you meet the candidate and give the check to him personally, if at all possible.   Never settle for less of a recipient than his campaign treasurer.

By substantial support we are talking about no less than $500 to a candidate for the House of Representatives and $1,000 or more to a Senate candidate.   It is understood that you take care of candidates from your own state first.   Be prepared to have your contribution listed in newspapers following the mandatory filing of reports for most campaigns.

A rather curious situation exists in certain states, known in political cant as "swing states."   These are states where neither party has enough of a following to consider the total vote safely Republican or safely Democratic.   The state thus swings back and forth between the parties.   Many congressional districts fall into the same category and House candidates running in swing districts can depend on reasonably generous assistance from their national party headquarters in general elections.   Especially in swing states it does not make a great deal of difference which of the parties you support on the state level, since work allotments and patronage are usually divided fairly evenly between the two major parties. In Pennsylvania, for example, all state design work is traditionally split down the middle and meted out by the respective political organizations to the party faithful.   Pennsylvania politicians further refine the work distribution by having a party chief responsible for each half of the state.   One man passes on all awards in the eastern half, another takes care of architectural and engineering contracts from Altoona west.

One can thus recognize the advisability of having principals active in both parties if there is any interest in garnering state design work.   A continually recurring problem in one large Philadelphia architectural firm, known far and wide for its Republican orientation, was finding a "house Democrat" to represent the firm at local, state, and national functions of that party. It was only after a re-organization of the firm into a partnership, in which one of the partners brought in from the outside turned out to be a Democrat, that proper coverage was assured in all political situations.

On the other hand, many states are so definitely and traditionally Republican or Democratic that to swim against the political tides is an exercise in futility. It would not be very productive of design work in the state of Missouri, for example, to advertise that you are a registered Republican.

THOSE PESKY DINNER TICKETS

When a firm has been the recipient of state or federal work many opportunities are afforded the firm's principals to attend political rallies, dinners, luncheons, and a wide variety of other social get-togethers, all aimed at replenishing party coffers. Tickets to these affairs range from $5 to $1,000 a plate. It has been observed that the more expensive the function, the more the pressure is to take a table, which never has fewer than ten seats. Elementary arith-metic shows that the desired attendance from a firm at a $1,000-per-plate dinner (desired by the sponsoring political group, that is) will set the principals back $10,000.

Now $10,000 is pretty important money to spend for the privilege of eating lobster with the President or whomever—even for the largest and most successful of design firms. Very few firms are given to making the full table gesture at those prices, although many take one or two tickets.

When the dinner host is the President or a Cabinet rank officer the managing partner or senior principal will usually be able to get away for the event. As the dinner prices range downward to the $25- and $10-a-plate city and country affairs it becomes in-creasingly difficult to find volunteers.

I discovered a trick of this particular trade some years ago. Most local and state party organizations are satisfied when two to four $25 tickets are bought to their "anniversary banquet" or "vic-tory dinner." (Losers seldom have celebration dinners.) Whether or not a representative of the firm shows up to drink watered bourbon, partake of the tired chicken, and sing thirty-two choruses of "For He's a Jolly Good Fellow" is usually immaterial. One can make an extra political point or two and avoid a boring, noisy evening at the same time by buying two or four or eight tickets, then turning them back to the party headquarters with the request that they be given to deserving state or city employees.   While it is not beyond the bounds of reason to think that such turned-back tickets are occasionally resold, they are usually distributed to employees of the sponsoring governmental unit who could not otherwise afford to attend—and who will definitely enjoy the opportunity to rub elbows with their higher-salaried superiors.   This tip alone should save more than the cost of the book in unneeded Alka-Seltzers, aspirin, and other stomach settlers and morning-after remedies.

UNPUBLICIZED CONTRIBUTIONS

To return to the subject of direct political contributions and the required listing of large contributors in the press and other media —there is a method wherein public identification as a major supporter of Senator Whosis or party X may be avoided.   This procedure is legal under existing election laws.   It also benefits from the fact that the contribution is even more gratefully received (and remembered) by the office seeker than the usual gift, the source of which immediately becomes public information.

The method is based on the premise that the winning candidate will have a campaign deficit.   (There is obviously no reason to consider this idea for losing candidates.)   Most campaigns do end up with a deficit these days, so that part should not be any problem.   Determine the deficit from news stories and other sources. If the amount is not too great and the official might be of definite future assistance, the grand gesture of covering the full amount of his campaign arrears might be made.   But let's assume the deficit is $25,000 or so.   Almost as effective as covering the whole amount is to offer to take care of 5 or 10 percent of the total.   On a $25,000 deficit he should not figure that you are trying to buy his soul and be most appreciative of the contribution.   Many congressmen become somewhat nervous over any contribution above $500, so know the man.

Up to now, the media has not seemed to have much interest in postcampaign contributions, presumably because a candidate and a campaign become old stories on the day after the election.   Under previous election laws, it was also a lot of trouble to track down late contributions. The real advantage of this type of deficit financing should be evident—you always back the winner.

PLAN CONTRIBUTIONS

In some architectural and engineering firms the principals make political contributions in predetermined amounts to certain key candidates. The amount of the contribution is covered by the firm in a concurrent advance, plus enough money in addition to cover the extra income taxes the individual making the contribution will incur. At the end of the fiscal year the bonus or profit-sharing checks are increased by the amount of the advance and the advance is then washed out in the year-end check. Obviously the firm is thus directing contributions to the candidates which its management wants to support—which is illegal in some cases. The morality of this procedure is left to the individual's conscience.

Prior to the enactment of the Federal Election Campaign Act in 1972, some corporations evaded the prohibition against corporate contributions by "loaning" personnel to a campaign or party headquarters. This was technically in violation of the 1925 Corrupt Practices Act, but such violations were apparently winked at by both political parties. In the 1972 campaign both parties insisted that any corporate employee take a formal leave of absence without pay from his firm before he would be accepted as a full-time campaign worker.

If, however, the employee maintained a reasonable semblance of keeping up with his regular duties and devoted something less than full time, plus evenings and weekends, to the campaign, the interpretation seemed to be that this was allowed under the Federal Election Campaign Act. In such cases the employee was not required to take an unpaid leave of absence. Corporations that have bonus and profit-sharing programs—and most do—were no doubt able to take care of any employees who were off salary for a month or two. Fortunate is the design firm with a political expert on its staff—for some way will be devised to utilize his talents.

DEDUCTIBLE CAMPAIGN GIFTS

As both major parties carefully explained in their many appeals for contributions to the 1972 campaign, every taxpayer may now take a small deduction or credit on his federal income taxes for most political contributions.

To qualify for the tax break a contributor should be certain that he is giving to a national political party or to a candidate actually running for a national, state, or local office. Some political action groups do not confine their activities to electing candidates, and contributions made to such organizations probably will not qualify for the deduction. Neither do contributed services or property qualify for the tax deduction. Large contributors may also be subject to federal gift taxes.

The Internal Revenue Service has ruled that the cost of a raffle ticket (in support of a party or a candidate) qualifies for the deduction only if the purchaser refuses the ticket and clearly makes an outright gift of the ticket price. Tickets to a political fund-raising luncheon or dinner are deductible if the meal is secondary to the political function. Otherwise, the value of the food must be subtracted from the cost of the ticket.

BUYING JOBS

Up to now we have dealt primarily with the making of contributions to candidates and parties in advance of elections—to help win an election. Those who know their way around state capitals and Washington, D.C., will occasionally hear about jobs "for sale." This usually means that for a contribution to the right party or individual of a flat percentage of the construction cost or the design fee the job can be "bought" if certain other considerations are met. This is probably as good a place as any to make the point that practically all opportunities to win a commission through political clout go to the larger firms which have an established record for good design and follow through. By the nature of things, these larger firms can afford to make significant campaign contributions. Politicians, by and large, are not inclined to award major jobs to five-man offices or to known design hacks.

ETHICAL POLITICAL ASSISTANCE

Since most professionals would rather win jobs on merit than through outright purchase, how may elected officials ethically be of assistance to architects, engineers, and planners? At whatever level of government—city, county, state, or national—political appointees and civil servants are normally responsive to inquiries from elected officials, whether the official be a city councilman, a county commissioner, a state legislator, or a member of Congress. Sometimes a call from a representative or a senator to a federal department for information about the progress of an award will be worth more to the design professional's interests than a dozen calls or letters from the designer himself. If his senator is on the committee which reviews that particular department's budget or appointees, so much the better for his chances, everything else being equal. A legislator can obtain copies of bills or laws on request or set up an appointment with a busy department head. Just don't overdo it; welcomes can be worn out.

For those who have not observed it first-hand, a telephone call from a congressman's office usually has a remarkable effect on the Washington bureaucracy. Some years ago I was trying to set up an appointment with Eugene Baughman, then head of the U.S. Secret Service. Several calls to Baughman went unreturned, which was understandable since he was a busy man. One day, while visiting the administrative assistant to a Midwestern congressman, I tried to contact Baughman and, as usual, he was not available. Since I would return to the congressional office after lunching with his aide, that number was left with Baughman's secretary. No mention was made that it was a House Office Building number.

By the time we got back to the office some two hours later there were three messages from Baughman, who had tried to return the call. Upon calling him again I was put through immediately and was able to set up the meeting for later that afternoon. Baughman's secretary was trained to respond to calls from "the Hill" — whether or not so identified by the caller.

An administrative assistant (AA) was mentioned above. These congressional alter egos can be of inestimable help in cutting through governmental red tape and obstinate officialdom. Around Washington, D.C. one top AA may be worth three run-of-the-mill congressmen in getting answers and action. Get acquainted with the AAs and other staff people in the offices of your representatives and senators.

WRITE YOUR CONGRESSMAN!

Public opinion — however it is expressed — is a powerful influence on legislative thinking. Tell your neighbors, your barber, or the milkman how you feel about state and national issues if you wish; but if you want results you must make yourself heard where things get done—in state capitals and in Washington, D.C. If you have gone to the trouble of helping to get a man elected to office you owe it to him to let him know your thinking on matters which come before him.

State your opinion or ask your question as succinctly as possible (state and national legislators *do* get a lot of mail). Avoid form letters furnished by organizations; they are readily identified as such by the receiver. A short note on one subject is best, a post-card is better than no communication at all, and occasionally a telegram commands the most attention.

U.S. and state senators may be addressed:

```
The Hon. John P. Doe
(or Sen. John P. Doe)
Senate Office Building
Washington, D.C. 20510
```

```
The Hon. John P. Doe
(or Sen. John P. Doe)
State House
Your State Capital
(or to his home address)
```

Dear Senator Doe:

Representatives, state and national, are addressed:

```
The Hon. John Jones
(or Rep. John Jones)
House Office Building
Washington, D.C. 20515
```

```
The Hon. John Jones
(or Rep. John Jones)
Your State Capital
(or to his home address)
```

Dear Mr. Jones:

Be polite in all correspondence with elected officials and you may expect their answers to be the same. If you expect an answer, let it be known. Many members of Congress use form answers to letters on certain issues. These will be Robotyped letters and therefore sometimes difficult of identification as form replies, but if you do not feel the answer is sufficiently to the point let your representative know. Eventually the stock replies will become exhausted and your letter will get personal and individual attention.

INFORMATION SOURCES

In addition to the reference sources listed in Chapter 4, including the *Congressional Staff Directory, Who's Who in American Politics,* and the *Congressional Directory,* there are several low-cost guides to U.S. and state legislators.   One of the better listings is the booklet published for each session of Congress by the Public Affairs Division of the National Association of Manufacturers, 1133 15th Street, N.W., Washington, D.C. 20005.   The NAM's 96-page *Who's Who in the 93rd Congress and the State Houses* sells for $2; for another quarter they will include the booklet, *Tips on Writing Your Congressman.*

REFERENCES

[1] By permission.   From *Webster's Third New International Dictionary* © 1971 by G. & C. Merriam Co., Publishers of the Merriam-Webster Dictionaries. P. 1326.

[2] © 1971 by the *New York Times* Company.   Reprinted by permission. Oct. 18, 1971, p. 31.

[3] Neil Hickey, "Election 72 — Make News, Not Commercials."   Reprinted with permission from TV GUIDE® Magazine.   Copyright © 1972 by Triangle Publications, Inc., Radnor, Pa., May 27, 1971.   Pp. 10 and 12.

[4] "Design Professionals Fix Bead on Elective Office," *Building Design & Construction,* Chicago, Ill. April 1972, pp. 35–37.

[5] From a talk delivered by Mrs. Mary Ellen Miller, Special Assistant to the Chairman, National Republican Congressional Committee, Denver, Col., April 3, 1971.

[6] © 1971 by the *New York Times* Company.   Reprinted by permission. Aug. 31, 1971, p. 47.

[7] David L. Rosenbloom, ed., *The Political Marketplace,* Quadrangle Books, New York, 1972, p. xiv.

[8] MEMO, Newsletter of the American Institute of Architects no. 437, Oct. 15, 1971, p. 2.

CHAPTER 10

Federal, State, and Municipal Clients

As many design firms have already discovered for themselves, working for governmental clients can be frustrating, interesting, and even challenging—and often all at the same time. A firm can lose a considerable amount of money on a project it never should have accepted, through a badly drawn contract or for many other reasons; or profits may be made which are comparable to those from some private jobs. It is perhaps this Mr. Hyde and Dr. Jekyll aspect of government work which attracts some firms and repels others out of hand.

GOVERNMENT AS A CLIENT

The average architect or consulting engineer, particularly the young practitioner, knows very little about contracting with the Federal Government. . . . A-Es [embark] on Government design work with the idea of producing an exemplary job only to find that their rewards were inadequate fees and telephone books of red tape.

The red tape will probably always be a part of Government work, so resign yourself to it. Realize from the outset that because of added paperwork, *your costs of producing a Government job may be higher than for a corresponding private project.*

If your fee is inadequate, you have only yourself to blame. Do not *take* a Government job if you cannot earn a fair fee. It is up to you to arrive at a just agreement with the Government Contracting Officer.[1]

On the other hand, be sure you understand the contract and know exactly what it is you are negotiating about. Government projects, like some in private industry, can be tricky and misleading. Not long ago a team of contract negotiators for a large A-E firm turned down a building for the Postal Service because they did not understand what they were negotiating for. They backed out of the multimillion dollar contract, they insisted, because the fee offered was less than 3 percent of the estimated value of the project. Unfortunately for their firm, the portion of the project that was being offered—on a flat fee basis—carried an effective fee of almost 6½ percent. The next firm in line snapped up the project, naturally.

The Private Practitioner, a joint publication of the AIA, the CEC, and the NSPE, suggests twelve "Dos and Don'ts" to keep in mind when pursuing federal architect and engineer work.

DOS AND DON'TS

1. Furnish each agency's Regional or Field Office with an up-to-date (not more than one year old) U.S. Government Architect-Engineer Questionnaire, Standard Form 251, in triplicate. Forms should be filed with the regional office in each area in which the A-E is interested in obtaining work. Blank forms are available at any agency, its Regional or Field Office, or at the AIA, CEC, or NSPE Washington offices. Form 251 may, and should, be supplemented with additional information, e.g., brochures, written articles, etc., illustrating experience and competence of the firm and its key personnel.

2. Become acquainted with each agency at the Regional Office level. Since many agencies have more than 1,000 Architect-Engineer Questionnaires on file, periodic visits will probably be necessary to acquaint agency personnel with individual firms and their qualifications. In addition to occasional visits, keep agency personnel apprised of design awards, technical articles, professional honors, or important promotions related to your firm.

3. Remember that selection of qualified architectural and engineering firms is most often made by selection committees which are set up within the agency and reviewed at the regional level.

4. Plan a visit to Washington to obtain general knowledge of pertinent agencies' operations. Make an appointment with your Washington professional society representative. He can help in arranging appointments with key Federal agency personnel.

5. While in Washington, include a visit to your Congressman's office to acquaint him with your firm and its interest in obtaining Federal work. An appointment should be obtained in advance and you should advise your Congressman if you are planning to discuss a specific project. Keep in mind that many Congressmen regularly furnish such guidance and assistance to constituents and this type of help is entirely ethical.

6. On a major project, you should expect to visit the agency's headquarters in Washington. The national offices normally retain the right to review and approve or disapprove the lists of qualified firms recommended by their Regional Offices or Field Offices. Consequently, personal knowledge by headquarters personnel of interested firms' qualifications can be helpful.

7. Be prepared to comply with Federal rules and regulations, e.g., audit of your contract, equal opportunity requirements, local wage determinations, adherence to Federal wage-hour laws, etc.

8. Don't try to obtain any project unless you are well qualified to perform it.

9. Exercise caution in employment of commission agents, who claim ability to obtain work by virtue of their government contacts. Under certain circumstances it is illegal to pay such persons a fee, commission, percentage or brokerage fee contingent upon, or resulting from, award of a given contract.

10. Subscribe to those Federal publications which are pertinent to your firm's area of interest. A source of project information is the *Commerce Business Daily,* available by subscription from Superintendent of Documents, Government Printing Office, Washington, D.C. 20402.

11. Don't be surprised if the first project with an agency may not be profitable. Know and understand your total obligations before you execute a contract.

12. Don't expect your efforts to lead you immediately to a commission. Competition is keen.[2]

Whatever a firm's past experience with government clients may have been, it will be more and more difficult to ignore government as a potential client in the future—particularly at the federal level. In a recent detailed analysis of its practice mix and client types a fairly large design firm found that governmental clients at all levels had accounted for 31 percent of its total business over a twenty-five-year span. This involved, incidentally, the design and construction of almost thirteen million square feet in a variety of structures. The government work total of approximately one-third of all its business is probably pretty representative of architects in general; engineering firms would find their figures to be somewhat higher, on the average.

In the business development study referred to above, the consultant set out certain findings and recommendations which are pertinent to this chapter:

> ... governmental clients [city, county, state, and national] have accounted for almost a third of our total work load. It would seem to be a pretty safe prediction that this figure will increase in the next 20 years—or at least the commissions and projects will increase and if we do not get our share it will be our own fault.

FEE CEILING PROBLEMS

One of the practical limitations to heavy involvement in government work has been the relatively low fee ceiling imposed. The federal government—or at least some of its departments and agencies—now seem to be more cooperative in getting around the six percent fee lid. Another innovation at the federal level holds some interesting future possibilities for firms geared to turning out turnkey projects. GSA and OFBO (Office of Foreign Building Operations) are two examples of agencies who have gotten into turnkey and pure lease-back deals; the OFBO because it apparently despairs of Congressional funding of needed chancery buildings and related overseas properties. In one way this approach might be considered a circumvention of Congressional intent, but this is not the concern of the design firm.

FEDERAL AID INCREASING

A 1971 study pointed out that federal aid to state and local governments in 1970 totaled almost $28 billion, as compared to some $8 billion in aid in 1960. In fiscal 1971 such aid totaled more than $32 billion—or four times the 1960 distribution. Federal funds accounted for some 20 percent of all local and state revenues in 1971; today the percentage has increased significantly.

Many of the points covered in these analyses are developed more fully in this and following chapters.

FORM 251

The first step in pursuing government work, assuming that a firm has decided it wants to enter or enlarge its hold on this field, is to complete the "U.S. Government Architect-Engineer Questionnaire—Standard Form 251." While submissions of Standard Form 251 are not required by many government bodies below the federal level, the form may be used in company with other brochures and promotional materials in any manner the design professional desires. Practically all departments, agencies, and bureaus of the federal government require the filing of a Form 251 before any consideration will be given to an architectural, engineering, or planning firm. A notable exception is the Agency for International Development, which has its own version, known, logically enough, as "Exception to S.F. 251."

Some of the pages in the AID exception vary considerably from their counterparts in Form 251, so if a firm is interested in AID projects there is no alternative but to follow that agency's format. AID requests that two sets of its form be filed with the Washington office. Contrary to the admonition in *The Private Practitioner's Dos and Don'ts,* it is not necessary to file more than one copy of the Form 251 with each government office in which you are interested. A number of state and municipal designer selection bodies use a questionnaire similar to Form 251. Unfortunately, because the format and order of questions is slightly different, there is nothing to do but fill in each one of these alien forms separately, even though the information requested is essentially the same in all cases. A worthwhile project for the national societies of architects, engineers, and other professional consultants would be to work with the national organizations of cities and counties toward the adoption of a universal A-E selection form—preferably the federal Form 251. Some other variations of Form 251, required by agencies primarily involved in funding foreign work, are illustrated and discussed in Chapter 12.

The Corps of Engineers requires a Form 251, but some of its district offices supply a supplementary form to be filled in as well. The second form is to expedite computerization of the essential information about a given firm. To cover all of the Corps of Engineers installations, including headquarters, divisions, districts, boards, commissions, and other activities, would require almost eighty copies of the 251 Form. Only the very large international design firms would have any reason to file with each installation, but this gives an idea of how far-flung some of our government agencies have become.

Get a supply of blank sets of Standard Form 251. They are available from government agencies, the headquarters of most professional societies, or in pads of twenty sets for $1 from the Government Printing Office in Washington, D.C. The basic set consists of nine numbered pages:

Page 1—General information, names of key personnel, and a breakdown of staff by occupation and location.

Page 2—Outside associates and consultants normally used and types of projects in which the firm specializes.

Pages 3 and 4—Personal history of principals and associates.

Page 5—Current projects for which the firm is architect or engineer of record.

Page 6—Current projects in which the firm is associated with others.

Page 7—Completed work (last ten years) for which the firm was architect or engineer of record.

Page 8—Completed work (last ten years) in which the firm associated with others.

Page 9—Security clearance information and certification of the information in the questionnaire.

Most firms need to insert extra pages in the sections on current and past projects in order to give complete answers. Large firms must usually add to pages 3 and 4, so as to have enough space to list all principals and associates. The completed Form 251 may run to 100 pages or more for a large firm in operation for many years or for an office which has been active in acquiring other firms.

MODIFYING FORM 251

As long as the basic pages are retained and the information requested is supplied there appears to be no limit to the adaptations a firm may make in its own Form 251. Many firms utilize special front and back covers to bring the exterior appearance of the 251 in line with the other elements of a brochure system. The inside pages and front and back cover are then assembled with a GBC or similar plastic binding along the top or left edge.

A cover for a Form 251 might be of colored stock and include the name of the submitting firm, the date the information was compiled, and a photograph of an outstanding project. If a photograph is used make sure it shows a federal job of some sort. This sounds elementary but one firm illustrated its Form 251 cover with a picture of an office building for a private client. A consulting firm features a map showing the location of its fourteen offices on the 251 cover.

Other modifications to the basic Form 251 format include a table of contents as the first page and dividers to separate the several inside sections. The dividers may be quarter or half sheets or full sheets with index tabs extending to the side.

Question 14 in Form 251 asks for an indication of the scope of services provided by a firm. Since the space allowed for the answer to this question is so small some 251s have two or more page inserts at this point (following page 2) to allow for a detailed explanation of the scope of services. A benday application to alternating boxes in the section for personal histories of principals and associates helps to break up a series of pretty deadly all-text pages.

Even though there is no prohibition against it, very few firms include photographs in their Form 251 brochures. Section 20 of Form 251 states:

> Unless specifically requested, submission of photographs is optional. Where submitted, furnish one exterior and one interior photograph of five examples of completed architectural work that are listed in items 18 and 19 (completed projects). Photographs of models, renderings, sketches, etc., are NOT desired. Size of photographs not to exceed $8\frac{1}{2}'' \times 11''$. On the back of each photograph give the following information: (1) Name of your firm; (2) Name and address of client; (3) Type of structure; (4) Location of structure; (5) Cost of specific structure. Photographs of electrical or mechanical facilities and other components of a decided engineering character are not necessary.

Some offices include a section of photographs and descriptions of a cross section of their work, emphasizing projects for federal agencies and departments. This technique also helps to break up the many pages of impersonal text. In short, an application of the principles of layout, graphics, and design to the Form 251 is certainly not forbidden, and a little applied imagination and ingenuity could help to set a firm's submission far above the usual Form 251.

While it says in the notes at the bottom of the last page of the 251 form that the form is to be completed by typewriter, a growing number of design firms use typeset fill-ins. Another approach is to have the blank form pages photostated up to twice size and then strip in the typed information. When the page is reduced back to its original size, if a good clear typewriter face was used and the ribbon was new enough to give a good black print, the reduced letter size should remain readable and be much neater in overall appearance. Once the Form 251 is filled in to everyone's satisfaction it should be reproduced in sufficient quantity to cover all of the federal agencies in which a firm is interested, plus an additional fifty to 100 copies. Xerox or similar copying processes may be used to reproduce the Form 251, but a much better job is obtained by having the pages run off on an offset press.

The extra copies are for use with state, county, and municipal bodies and for all types of other clients: commercial, medical, educational, etc. The Form 251 works as an excellent supplement to the firm's regular brochure, in that it gives a potential client a quick review in condensed form of the staff, experience, recently completed work, and other data which may be of interest.

The discussion thus far has been concerned with getting information copies of a design firm's Form 251 on file in as many government offices as the principals of the firm believe is practical. In addition to the mass-produced, mass-distributed version of Form 251, which must be on file in order for a firm to receive consideration by a government agency, some firms make it a point to file a special 251 in connection with their interest in a specific project. It is prudent, for example, to file a customized Form 251 for a highly specialized, extremely complex, or particularly large project. When two or more firms plan an association or joint venture for an upcoming job, they should file a revised Form 251 to indicate their joint status.

ADVICE FROM THE EXPERTS

Representatives of government contracting offices generally agree on several suggestions to A-Es for completing Standard Form 251.

1. On the pages for listing current and completed projects, break down the list into building or client types, e.g., office buildings; educational facilities; medical, clinic, and rehabilitation facilities; commercial buildings; research and production facilities; public buildings; industrial buildings and master planning. The Veterans Administration, for example, will be more interested in a firm's experience with hospitals, clinics, nursing homes, and rehabilitation facilities than any other types of projects.

2. Consider the needs and special interests of the potential client — the government agency — carefully; then customize the Form 251 to whatever extent might be appropriate. As in the VA example above, move the pages listing medical and medical-related jobs to the beginning of the 251 sections on current and past projects.

3. View the 251 as any application to seek employment. For an important position, one does not send a mass-produced resume; rather it is customized and slanted to fit the employer's known requirements. And woe to the applicant who does not trouble himself to learn as much as possible about a prospective employer before submitting an application.

4. If a firm has an extremely long list of projects under one building type heading (particularly if it is a building type unrelated to a specific project being sought) list only a dozen or so representative projects. Make it clear that they are only representative.

5. Do not forget to show totals of projects and construction costs at the bottom of each page. If the projects are categorized, as recommended above, then show totals at the end of each category and a grand total at the end of the sections on current and past work. Someone has to provide these totals; if it is the potential government client, then the A-E has not made a very good initial impression on him.

6. Short sentences and short paragraphs are one of the keys to readability in regular prose writing, so look for ways in the job listings to break up the monotony of page after page of single spaced entries. One simple method is to skip a space after each group of five or ten entries.

7. Remember to fill in the date the form was compiled in the box in the top right-hand corner of page one. Another often-missed item is the date and signature required in question 23 on the last page of Form 251.

8. If photographs are included in or with the 251, be certain that they are relevant and "show the architecture." Government selection bodies are not interested in viewing photographic tricks or stunts, wherein the appearance of the completed building has been modified to satisfy a whim of the designer or his photographer. Include several exterior views of a building to show the significant elevations.

9. One last piece of advice from the government representatives —which has application to almost any proposal or presentation from a designer. The A-E should always ask himself what it is he can do for a client which no one else can do. The design professional must offer some special incentive (Unique Selling Proposition) to the potential client to justify his being chosen over six to twelve competitors all of about the same size and experience in the case of government work. Then put down the information

to explain this incentive or set of advantages in a manner at once easily read, comprehensive, and immediately digestible, so that it will have the desired impact and effect on the potential client.

WHERE TO FILE FORM 251

The 251 must be placed with all federal agencies from which commissions are desired. Do not depend upon one office of an agency to distribute copies of your 251 to the other offices of that agency in a region or area—make the distribution yourself to insure complete coverage. Aim for saturation; if an office ends up with more than one copy of a firm's 251 the extra ones can always be discarded, but if no copy is received then that design firm does not exist as far as the federal agency is concerned.

If a firm has branch offices in other areas or is capable of regional or national operations, copies of Standard Form 251 should be placed with the Washington, D.C., offices of all federal departments, bureaus, and agencies.

The Form 251 must be revised periodically. Some agencies, because of storage limitations, make it a practice to throw out all 251s after a period of time—usually one to three years. It is advisable to update the 251 every six months. This keeps the file current and every new edition is an excuse for a brief contact, either in person or by letter, with federal contracting officers. No more than twelve months should ever lapse between submissions of Form 251. As a reminder to themselves and to the federal office involved some design firms put the date of the current 251 in a prominent place on the cover.

A list of government offices for placement of Form 251 will be found later in this chapter.

WASHINGTON REPRESENTATION

Most professional design offices receive periodic mailings from one or more of the "Washington reps." These are consulting firms located in the District of Columbia who, for a monthly retainer, offer varying degrees of representation, specialization, experience, contacts, and influence. Fees for this type of service can range from $1,000 a year to several thousand dollars a month.

Some Washington representatives provide a valid service; others somehow maintain staffs and plush offices on a combination of unkept promises, undelivered influence, hot air, and sheer gall. It is the latter group, of course, that the design professional must somehow avoid if he really believes a representative in Washington would be beneficial to his practice.   In an editorial about the famous purloined IT&T memo in March 1972, the *Washington Post* took an unkind cut at the entire Washington rep corps: ". . . [their] overblown claims of influence effectively brought to bear, which are the stock in trade of the successful lobbyist. . . ."

All Washington lobbyists or representatives base their sales approach and fees on the premise that few laymen are qualified to find their own way through the quagmire of red tape, bureaucratic indifference and ineptitude, and sundry other small- and large-scale obstacles on the Washington scene.   While this premise is not necessarily false, the concomitant assumption that just any lobbyist is better equipped than the neophyte to serve as a guide to the seats of influence and power is demonstrably in error.

Where a legitimate advisory and information service is provided by letter and telephone, accompanied by personal consultation and assistance when the client visits Washington, D.C., few would quarrel with the idea that this is worth a reasonable retainer.

If you believe that Washington representation might be a good idea for your firm there are a number of ways to learn who are the effective, professionally oriented representatives.   Staff members of professional societies headquartered in Washington (as most are) should be able to come up with some names of legitimate counseling firms.   Your representatives in Congress, or their staff members, certainly know the Washington lobbyists—good and bad—and if your political fences are in order you might get a few good leads from the Hill.   Fellow professionals, who already have Washington representation, may be willing to provide the name of their rep.   This is not to suggest that you consider retaining another design firm's representative; his own ethics should normally preclude his taking on competing clients and if this is not the case, any design professional should be astute enough to avoid such a potential conflict of interest.   There are, however, a few good Washington-based counseling firms which represent several regional professional clients, such as architects and engineers, and have little, if any, conflict of interest to be concerned about if they take care to avoid overlapping regions and duplication of Washington interests.

What are some of the services one might expect from a Washington representative? The following excerpts from a solicitation letter written by a knowledgeable and experienced Washington rep should be of some guidance.

The basic role of the Washington office would be to provide the following services:

1. To provide up-to-date information on all pending and available federal capital improvement grant programs, the agencies administering them, and their requirements.

2. Management talent and time and ability to:

(a) Study the applicability of particular aid programs to local needs.

(b) Aid in the preparation of acceptable applications for municipal clients for capital grant requests.

(c) Follow through on them promptly and effectively at the Washington and regional levels.

3. To serve as a single information source for all types of federal, state and private assistance emanating from Washington, D.C.

4. To develop good working relationships with federal program administrators within Washington agencies and maintain close contact with appropriate Senators and Representatives.

5. To review weekly reports, surveys and studies to determine prospective clients for architectural grant-in-aid assistance, noting their requirements and potential benefits and communicating this information to the client. In this way an "early warning system" would enable the firm to be ready to utilize federal programs to "create" municipal clients when programs are finally established and funded by Congress.

6. To provide an assembly and coordinating center for program information related to implementing planning contracts in which the firm may be engaged. Each planning summary could include programs which would aid the county or city to proceed into the implementation phase. Such information would include, but not be limited to, the following categories:

(a) Program description (the nature and purpose of the program and availability of more detailed information).

(b) Administering federal agency.

(c) Types of assistance available (grant, loan, technical assistance).

(d) Eligibility (who can apply and who can benefit).

(e) Functional purpose for which money can be spent (planning, research, construction, training).

(f) General conditions for receiving aid (comprehensive plan, state plan, matching funds).

(g) Funding (authorizations, appropriations, regional allocations, remaining funds available).

(h) Related programs administered by the same agency.

(i) Related programs administered by other federal agencies.

(j)   Application information (where to get forms, deadlines).

(k)   Authorizing legislation (public law, number and title).

(l)   Administrative requirements (legal authority for participation, federal and state reporting and evaluation requirements, auditing).

(m) Processing time (average to approve, reject or return an application).

Perhaps the most important information the office could supply a prospective or actual municipal client are answers to the following questions:

1. Does the funding agency have money for this program?

2. If so, how much?

3. If not, when will money be available?

4. How many applicants are there?

5. Are pre-application guidelines available?

6. How are funds to be allocated within the region and within the state?

7. Where are the general priorities within the program?

8. What is the general processing time?

This is the type of information usually omitted by consulting firms on general planning contracts, but it is the most important input which the municipal client requires.

This proposal suggests the provision of a variety of services; to the prime client (a professional design firm) and, by extension, to the firm's own local and state clients.   Another Washington-based organization described its "Information and Advisory Service" in these terms:

This covers—immediately as legislation is considered and/or authorized —specific data on:

Fully Funded Construction Projects in your designated area of interest.   Such data will include: GSA prospectuses as required under the Public Buildings Act of 1959 on major Federal buildings; itemization of planned construction by VA, HEW, NASA, AEC, Agriculture, State, etc.; guidance on Department of Defense and Public Works projects.

Advice as to responsibility for consideration, selection and contracting of A-E services for the projects indicated and instructions as to implementing procedures.

Partially Funded Construction Programs providing for construction grants and/or loans to states, counties, cities and community groups in their expansion of civic, educational, medical, social and other programs.   Examples: Demonstration Cities and Metropolitan Development Act of 1966; 1966 Amendments of Higher Education Facilities Acts; Elementary and Secondary Education Act; Library Services and Construction Act; Urban Mass Transportation Act; etc.

Projects involved must be initiated by state, interstate or local government entities or by civic groups in *association with locally selected A-E
firms.* Advice as to required implementation will be furnished.

CONTINGENCY FEES

A service for which one should *not* look to his Washington representative is that of finding work on a fee contingency basis. (See
number 9 under the Washington "Dos and Don'ts" earlier in this
chapter.) One of the many forms a firm is required to sign before
a contract can be executed for federal design work is the two-page
"Representations and Certifications." Certification number 2
covers contingent fees and the design professional must answer
whether or not he has ". . . employed or retained any company or
person (other than a full-time bona fide employee working solely
for the A-E) to solicit or secure this contract . . . ," and if he has
". . . paid or agreed to pay any company or person (other than a
full-time bona fide employee working solely for the A-E) any fee,
commission, percentage or brokerage fee, contingent upon or resulting from the award of this contract . . ." The A-E agrees to
furnish all information relating to these two questions to the
contracting officer upon request. The penalties for withholding
or supplying false information include fines and prison.

One firm of Washington representatives derives the greater part
of its income from collecting 8 to 10 percent of the A-E's gross
fee for all federal projects it claims responsibility for placing with
its design clients. Some clients pay a stiff monthly retainer in
addition to the percentage of design fees. To my knowledge, no
registered design professionals head any of the fee-splitting
Washington reps, some of which are really little more than influence peddlers.

D.C. BRANCH OFFICE

The alternative to engaging a Washington representative, if a firm
believes its best interests will be served by having someone in
residence, is to establish a Washington, D.C., branch office. Many
firms have done so and the trend appears to be up.

All the usual considerations already discussed for opening a
branch office apply in this case. Washington branch offices vary
all the way from a storefront operation with a secretary to forward

mail and handle telephone calls, to a full production office offering complete design services and a wide range of support services, including marketing. The one-girl offices do not impress most federal agencies, in that they are rather obvious fronts not intended to produce anything within the District. The best to be said for this type of office is that it is usually—but not always—better than no office at all.

FEDERAL CLIENTS

More than 100 agencies, bureaus, and administrations contract for architect-engineer services on a regular to infrequent basis. Many of these offices also maintain regional or state offices, and the design firm in serious pursuit of federal work will see that copies of its Form 251 are filed in every possible outlet. Addresses of individual regional offices may be obtained from local telephone directories; lists of all external offices of an agency are available from its headquarters—most of which are located in Washington, D.C. A partial listing of the agencies most apt to require architectural, engineering, and planning services follows:

INTERNATIONAL DEVELOPMENT

    Agency for International Development (AID)
    Inter-American Development Bank (IDB)
    Export-Import Bank of the United States (Ex-Im Bank)
    Overseas Private Investment Corporation (OPIC)
    International Bank for Reconstruction and Development (World Bank)
    International Development Association (IDA)
    UN Development Program (UNDP)
    U.S. Treasury Department
      Office of Assistant Secretary for International
      Affairs
    U.S. Department of Commerce
      Bureau of International Commerce
      Office of International Business Assistance
      Export Business Relations Division
    U.S. Department of State
      Deputy Assistant Secretary for Commerce and Business Activities
      International Business Affairs Division
      Office of Foreign Buildings

DEPARTMENTS

    Agriculture, Department of
      Farmers Home Administration

Forest Service
Rural Electrification Administration
Commerce, Department of
Economic Development Administration
Maritime Administration
National Bureau of Standards
Defense, Department of
Air Force
Director of Civil Engineering
Army
Office of Chief of Engineers, Director of
Military Construction
Civil Defense, Office of
Navy
Naval Facilities Engineering Command
Health, Education and Welfare, Department of
Facilities Engineering and Construction Agency
Housing and Urban Development, Department of
Federal Housing Administration
Community Planning and Management
Community Development
Housing Management
Research and Technology
Interior, Department of
Alaska Power Administration
Bonneville Power Administration
Bureau of Indian Affairs
Bureau of Land Management
Bureau of Outdoor Recreation
Bureau of Reclamation
Delaware River Basin Commission
Fish and Wildlife Service
Geological Survey
National Park Service
Office of Saline Water
Justice, Department of
Bureau of Prisons
Law Enforcement Administration Agency
Labor, Department of

U.S. Postal Service
  Corps of Engineers
State, Department of
  Bureau of Technical Assistance
  Office of Foreign Buildings
Transportation, Department of
  Federal Aviation Administration
  Federal Railroad Administration
  U.S. Coast Guard
  Urban Mass Transportation Administration
Treasury, Department of

AGENCIES

Atomic Energy Commission
Environmental Protection Agency
  Solid Wastes Management Programs
  Categorical Programs
  Air Programs
General Services Administration
  Public Building Service
National Aeronautics and Space Administration
National Water Commission
Tennessee Valley Authority
United States Information Agency
Veterans Administration
Water Resources Council

This list is correct as of September 1971. Many of the above departments and agencies have district and regional offices concerned with design and construction. These include:

Army Corps of Engineers
Naval Facilities Engineering Command
Air Force
Health, Education and Welfare, Department of
General Services Administration

Certain other offices, such as the Army and Air Force Exchange, are difficult to classify, but have responsibility for design and construction.

ROLE OF P.L. 92-582

P.L. 92-582, perhaps better known in the profession as the "Brooks bill," after one of its sponsors, Texas Representative Jack Brooks, is an act "to amend the Federal Property and Administrative Services Act of 1949 in order to establish Federal policy concerning the selection of firms and individuals to perform architectural, engineering and related services for the Federal Government." The Brooks bill became a law of the land on October 27, 1972, and has some highly important and significant sections for design professionals.

We will not take the time here to go into the long history of P.L. 92-582 and the many attempts to get its forerunners through Congress. Suffice it to say that it was finally passed and signed into law after some unbelievable delays and over formidable obstacles thrown up by various individuals and departments in the executive and legislative branches.

The three pertinent sections of P.L. 92-582 are

POLICY

SEC. 902.   The Congress hereby declares it to be the policy of the Federal Government to publicly announce all requirements for architectural and engineering services, and to negotiate contracts for architectural and engineering services on the basis of demonstrated competence and qualification for the type of professional services required and at fair and reasonable prices.

REQUESTS FOR DATA ON ARCHITECTURAL AND ENGINEERING SERVICES

SEC. 903.   In the procurement of architectural and engineering services, the agency head shall encourage firms engaged in the lawful practice of their profession to submit annually a statement of qualifications and performance data. The agency head, for each proposed project, shall evaluate current statements of qualifications and performance data on file with the agency, together with those that may be submitted by other firms regarding the proposed project, and shall conduct discussions with no less than three firms regarding anticipated concepts and the relative utility of alternative methods of approach for furnishing the required services and then shall select therefrom, in order of preference, based upon criteria established and published by him, no less than three of the firms deemed to be the most highly qualified to provide the services required.

NEGOTIATION OF CONTRACTS FOR ARCHITECTURAL AND ENGINEERING SERVICES

SEC. 904. (*a*) The agency head shall negotiate a contract with the highest qualified firm for architectural and engineering services at compensation which the agency head determines is fair and reasonable to the Government. In making such determination, the agency head shall take into account the estimated value of the services to be rendered, the scope, complexity, and professional nature thereof.

(*b*) Should the agency head be unable to negotiate a satisfactory contract with the firm considered to be the most qualified, at a price he determines to be fair and reasonable to the Government, negotiations with that firm should be formally terminated. The agency head should then undertake negotiations with the second most qualified firm. Failing accord with the second most qualified firm, the agency head should terminate negotiations. The agency head should then undertake negotiations with the third most qualified firm.

(*c*) Should the agency head be unable to negotiate a satisfactory contract with any of the selected firms, he shall select additional firms in order of their competence and qualification and continue negotiations in accordance with this section until an agreement is reached.

Full implementation of Section 902, concerning public announcement of "all requirements for architectural and engineering services," was still being worked out as of this writing, but for the first time many federal departments and agencies are now listing projects in the *Commerce Business Daily* and the *Daily Federal Register*. A definitive and authoritative interpretation of what should go into the public announcement and where it must be placed is still being developed.

Section 903 spells out a formal requirement for A-E firms "to submit annually a statement of qualifications and performance data." For better or for worse, this statement will be the U.S. Government Architect-Engineer Questionnaire—Standard Form 251, until and unless superseded. Section 903 also seems to call for formal interviews with at least three A-E firms by, or on behalf of, each federal agency head for every project requiring architectural, engineering, or related services. We say "seems" because an official interpretation of the phrase is yet to be formularized. Where a project will be designed by an agency's in-house staff, it obviously would be superfluous to have presentations made by outside, independent firms. Some minimum cutoff figure, such as for all projects valued at $10,000 or less, is also being sought, to avoid having to advertise and interview firms for small remodel work and insignificant additions — most of which would be done in-house at any rate.

OTHER REFERENCE SOURCES

In addition to some of the reference books listed in Chapter 4 (*Congressional Directory, U.S. Government Organization Manual*, and the *Congressional Staff Directory*) there are other helpful directories available to assist those trying to chart a path to federal jobs. Two of these are:

Directory of Key Government Personnel: A publication of Hill and Knowlton, Inc., 1 McPherson Square, N.W., Washington, D.C. 20005. An excellent short guide but subject to outdating. Free.

Congressional Handbook: A publication of the Chamber of Commerce of the United States, 1615 H St., N.W., Washington, D.C. 20006. Not as complete as the Hill and Knowlton booklet, but it is not subject to outdating since a new handbook is published for each session of Congress. Single copies 50 cents.

PURSUING STATE AND
COUNTY WORK

Much of the preceding material on dealing with the federal government has direct application to business development efforts in a firm's home state and county. An increasing amount of federal money is allocated to various state agencies for spending. The Law Enforcement Administration Agency of the Department of Justice is one example of an agency which channels most of its funds through state agencies dealing with crime and detention facilities.

Most of the larger state agencies have regional or county office branches, so it is important for a firm to be known at all levels. Officials and staff members of the state agency or commission which is responsible for luring new business into the state are good contacts, as was pointed out in an earlier chapter.

Unless the project is very large or requires a hard-to-find design specialty, it is difficult for most firms to find work on their own from other than their own state governments. The association or joint venture between out-of-state and instate offices is one way of surmounting such jingoistic barriers. One can always find rare exceptions to this—and the exceptions, in addition to proving the rule, are often a revealing breath of fresh air on the design scene. One such exception—and there are others—is the University of Massachusetts, with its three campuses scattered across the state. On the Amherst campus, for example, one finds buildings designed by many of the United States' best known design firms, selected without regard to their state of residence. This open-mindedness is due, in large part, to the university's planning department, backed up by the administration and the building and grounds committee.

One inevitably will find political considerations and pressures creeping in on jobs from most state agencies. These considerations and pressures intensify, in most cases, as the architect and engineer move down through county and municipal levels and, incidentally, closer to home. Fortunately, there are also some notable exceptions to this generality.

MUNICIPAL WORK

Some not-so-notable exceptions to the above generality are the large U.S. cities such as New York, Philadelphia, and Chicago. Not only does the design professional encounter political considerations when he does finally win a commission, but also he must often deal with archaic regulations, underpaid and underqualified employees, lower-than-average fee schedules, slow payments, and systems of built-in delays which can stretch design and construction time over periods of up to ten years in extreme cases.

New York is not an average city in any sense of the word, of course, so it not surprising that a 1970–1971 Urban Design Council study found that "the city receives the poorest design at the greatest cost and with the longest delays through building processes." One of the most depressing aspects of the situation is that municipal construction accounts for one-third of the total construction dollar in New York City: an estimated $15 billion in the decade of 1970 to 1980.

One of the city's major problems is the Multiple Contracts Law, requiring separate bids for electrical, plumbing, mechanical, and general contracts. Enacted originally to reduce costs, the result has been to drive them upward and to compound the usual on-site chaos.

Concurrent with the release of the Urban Design Council's scathing report, the *New York Times* carried an article, "City's Building Process: How It Works," which is illustrative of the outer limits of poor regulations, worse administration, and municipal procrastination:

A city agency that needs a new building is required to prepare an "initial project program" and budget estimate for the City Planning Commission 10 months before the capital budget is certified by the Mayor and 15 months before the agency can expect construction to begin. Actual time elapsed before construction starts is usually two to three years.

The Planning Commission holds draft capital budget hearings. Based on the draft capital budget, the Bureau of the Budget produces the executive capital budget. The executive capital budget goes to the Controller and the Planning Commission. Public hearings are held by the Board of Estimate and the City Council.

The construction item, purposely vague, can be modified or eliminated anywhere along the way. The board or the Council may change or omit it. The Mayor can veto the change. A two-thirds vote of both the board and the Council is needed to reinstate it.

The budget is approved. The agency receives authorization to initiate the building project. The agency suggests a site to the Site Selection Board. The board holds public hearings. The site is approved.

The agency, working with the Municipal Services Administration, prepares a more detailed building program, within the approved budget. This becomes part of the architect's contract. The architect is selected from the Mayor's list of approved firms. He is handed the program, the site, the city's Basic Services Contract and the standard fee schedule. He accepts or rejects everything on a take-it-or-leave-it basis.

The architect's fee, 4.5 percent to 6.5 percent, depending on cost and complexity, is lower than most cities and American Institute of Architects schedules provide. It is based on conjectural preliminary cost estimates, an uncommon practice. Realistic adjustments are permitted within limits.

The architect's work is spelled out in five phases, with part payment on each phase. The largest emphasis is on construction documents; the smallest is on design.

During design, there are reviews by the agency, the Bureau of the Budget, the Art Commission and the Controller's office. There may also be reviews from the Transit Authority, the Environmental Protection Administration, Fire and Police Departments, the Board of Standards and Appeals, or the Board of Health. The project is approved.

Construction bids are invited. Under the Multiple Contracts Law, separate bids are required for plumbing, electrical, mechanical and general contractors.

Because the architect's contract does not provide for him to supervise construction, as in private practice, additional documents are required, detailing construction coordination of the multiple contractors. Uncoordinated change orders come from the separate contractors. The city supervises with short manpower.

The building is finished. Costs have risen 10 to 12 percent per year during the lengthy process. Post-audits of the Controller's office hold up payments. The architect must often borrow money at higher interest rates than his profit rate.[3]

The Urban Design Council of New York called this process "a Kafka-like obstacle race to excellence or economy."

SPECIALIZED CONSULTING

Municipalities and transportation authorities may be potential clients for a little-known and somewhat specialized type of consulting work from architects and engineers. Projects funded by bond issues usually provide for periodic inspection and evaluation of the completed facilities as a protection for the bond holders. These inspections may be done by the original designers, but are not necessarily restricted to them. A Public Notice from the Jacksonville, Florida, Port Authority in late 1971 illustrates how this type of work is obtained:

PUBLIC NOTICE

The Jacksonville Port Authority proposes to employ a nationally known and recognized firm of architects-engineers to inspect the airports and aviation facilities of the Authority and to make a report thereon and recommendations relating thereto in accordance with Ordinance No. FF 253, Bill NO. FF 296, authorizing issuance of Airport General Improvement and Revenue Bonds, issue of 1965.

Interested parties should contact the Director of Engineering, Jacksonville Port Authority, 19th Street and Talleyrand Avenue (telephone 356-1971) to receive further information pertaining to this project no later than August 9, 1971.

The authorization, in this case, is to be found in the ordinance authorizing the employment of the architect to design the original airport, Article IV, Section 12:

ARCHITECTS-ENGINEERS AND AIRPORT CONSULTANTS. That the City will employ Reynolds, Smith and Hills, Architects-Engineers, Jacksonville,

Florida, or other nationally known and recognized architects-engineers, to supervise the construction of said new airport and aviation facilities provided for herein and to inspect said airports and aviation facilities of the City at least once in every three-year period, and at least once in each year if requested to do so in writing by the holders of not less than twenty-five per centum in aggregate principal amount of said Bonds then outstanding, and to make a report thereon and recommendations relating thereto.

This is the type of job which, for various reasons, is not often publicized to any great degree. It may require some digging into old ordinances to learn what is required of local authorities and commissions in connection with local bond issues.

 Another type of bread-and-butter activity, not connected with municipal work, is the review of residential building plans for mortgagors such as savings and loan associations. Louisiana is one state that by law requires a professional review of plans. A few firms in New Orleans take care of a lot of their overhead by the fees from such work.

PAYOFFS AND KICKBACKS

This touchy subject has already been covered to some extent in other chapters, such as the one on "Political Action." It is a fact of life that some county and municipal officials expect part of the design fee routinely to find its way back into local party coffers or into individual pockets. Whenever and wherever these arrangements are detected and the culprits punished, the resulting publicity does not seem to affect the system to any marked degree.

There are more insidious avenues of graft open to venal public officials than straight-cash kickbacks. One of these is the requirement, made clear in advance, that the government unit client will name all consultants. In this manner the lead firm of an association may avoid direct payoffs, but the principal should be under no misunderstanding about the role of his consultants.

It is hardly enough to gloss over such matters with "everybody does it"—because "everybody" doesn't do it. And, while it appears that many more get away with it than get caught, the attendant publicity for a design firm that is caught can be extremely injurious professionally.

REFERENCES

¹ Gilbert A. Cuneo et al. Material excerpted from *Contracting with the Federal Government,* Washington, D.C., 1969, with permission of the American Institute of Architects.   Pp. 1–2.

² *The Private Practitioner—Securing of Federal Architect-Engineer Work.* Quoted with permission of the American Institute of Architects, Washington, D.C., 1971.

³ © 1971 by the *New York Times* Company.   Reprinted by permission.   Nov. 23, 1971, p. 45.

CHAPTER 11

Joint Ventures, Associations, and Other Consortia

Several references to associations and joint ventures have been made previously in these pages. In Chapter 4, for example, we pointed out that other professionals are frequent sources of job leads and "are usually potential associates or joint venturers, who, for various reasons, do not feel that they can obtain or produce the job by themselves. Perhaps their firm is too small or too busy, or the job is too complex or the potential client has specifically asked for a certain kind of expertise they do not have. Whatever the reason, you will probably be approached from time to time about entering into an association for a specific job."

DEFINITIONS

We need not concern ourselves very much at this point with whether the professional amalgamation is called an association, a joint venture, or a consortium. The differences are minor in practice and turn on legal definitions and liability insurance considerations for the most part. A few legal and lay definitions of the terms will either help your understanding or further confuse you. It is understood that the design professional will always have professional legal assistance in drawing up all contracts.

Judge Bernard Tomson and attorney Norman Coplan, in *Architectural & Engineering Law*, explain the difference between a partnership and a joint venture:

> Two or more persons or firms may associate together to carry on a single specific enterprise, such as the performance of a contract with a third party to render architectural or engineering services for a single construction project. This type of association, closely akin to the partnership, is the joint venture. The only distinction between the two is that a joint venture is formed to carry on a single specified enterprise or transaction, whereas a partnership is formed for the transaction of a general and continuing business of a particular kind.[1]

Justin Sweet, in his recent book on legal considerations for the design and construction fields, explains several meanings of the term "association." Pertinent to this discussion is this passage:

> [One] use of "associate" relates to the arrangement that may be made between two architectural firms to perform certain work. For example, a New York firm might associate with a firm in Los Angeles, if the New York firm has agreed to design and administer a project in Los Angeles. There may be many reasons to "associate" with a local firm. Such an association could be a joint venture if the firms agree to share profits and losses in some manner. If the local firm merely agrees to perform certain designated functions, without any stake in the profits, the local firm is merely performing these services on a contracting, rather than an entrepreneurial, basis.
>
> Similar to the "association" with a local firm is the arrangement under which an architect "associates" with another local architect or engineer. The purpose of this association may be to use the particular skill of the other architect or engineer. This is common in complex construction projects. Unless there is a true joint venture, this form of "association" is like hiring a local firm for an out-of-town project. Such "associations" are unlike the other associations discussed, since there is no element of joining together in an entrepreneurial sense.[2]

"Adventure" or "joint adventure" are proper legalese, incidentally, for joint venture. This terminology dates back to early maritime law, when associating to ship or bring back cargoes from overseas ports was literally an "adventure" for all concerned.

One final definition from our legal brethren is found in *Black's Law Dictionary*. If you are not familiar with Henry Campbell Black's exhaustive listing of legal terms, you should be. It won't take the place of your attorney but *Black's* can be helpful in translating some of the things attorneys write and say. Black quotes

Griffin v. Reilly, Tex. Civ. App., 275 S.W. 242, 246, on "joint adventures": "A special combination of two or more persons, where in some specific adventure a profit is jointly sought, without any actual partnership or corporate designation.[3]

The definitions of associations, joint ventures, and consortia will conclude with a few words on each from *Webster's Third New International Dictionary:*

> Association: an organization of persons having a common interest; a body of persons organized for the prosecution of some purpose, having no charter from the state, but having the general form and mode of procedure of a corporation.[4]

> Joint venture: a partnership or cooperative agreement between two or more persons restricted to a single specific undertaking.[5]

> Consortium: an international business or banking agreement or combination (as for the financial assistance of another nation or for the control of a particular industry in a country or countries).[6]

WHY JOINT VENTURES

All of this may sound overly complicated to the uninitiated. One might ask why the client does not simply retain one firm with the necessary expertise and background and sufficient experienced staff to accomplish the commission. Frankly, we are talking here primarily about clients who are municipal, county, or state political entities, represented by appointed or elected members who serve in one way or another at the pleasure of a constituency. This also applies in varying degrees to federal projects, as we have seen, and to overseas jobs, covered in Chapter 12.

Whatever quarrel one might have with parochialism and provincialism, the practicalities of the situation often dictate that local design firms must be involved. Backing up their politicians, local chapters of the American Institute of Architects and the professional engineering societies frequently issue statements about "carpetbaggers" invading their territory, taking hard-to-come-by municipal and county tax funds out of the community. Interestingly enough, some of the loudest complainers about "foreign" design firms are the so-called national offices, whose continuing growth and livelihood depend on their obtaining jobs all over the United States, if not from all over the world.

Chicago may be the leading example of a major metropolitan area excluding outside design firms, for all practical purposes, while its locally based design firms range from one corner of the globe to another in search of work. This certainly applies to public undertakings (city and county), but private work seems to run into much the same kind of restrictions. Ed Stone, to the best of my knowledge, was the first major designer from outside metropolitan Chicago to crack the scene, when he was joined with Perkins & Will on the Standard of Indiana office building. One cannot get too upset about the situation in Chicago since so many of the nation's top design firms are located there, but it does serve as a big city example of organized but largely unofficial efforts to shut out "foreign" designers.

The same obstacles, adjusted to local conditions and mores, are present in the great majority of governmental units throughout the United States. Recently, in a large city in the Midwest (not Chicago) the city fathers determined it was time to turn the old city auditorium into a full-fledged convention center, with a new arena and greatly expanded exhibition areas. One member of the city council had the bad judgment to state publicly that he wanted to see the city retain the "best architects and engineers in the world" to design the new facility. The project was estimated to be worth about $25 million and this happened to be a city hit by more than its share of strikes in the construction industry. Every local design firm left in business was extremely hungry. When the local chauvinists finished raking the hapless councilman over the coals, he was only too happy to get up at the next council meeting and make it perfectly clear that he really meant the "best architects and engineers in the world in *our* city."

The ironic postscript to this story is that the job finally went to a Chicago firm, in association with local firms.

We are not suggesting that the jingoists be driven from the city halls, county courthouse buildings, and state capitals of our land; the tradition of "buy at home" is undoubtedly too well established to be effectively uprooted now. As we saw in the last chapter, even federal job awards — particularly those of lesser importance and cost — are subject to the same considerations of local residence. As for major federally funded projects worth $2 million or so and up, rarely are they awarded to a recognized national firm without at least one associate office from the local area being part of the design team. Only when a job reaches upwards of $50 million or so does provincialism sometimes take a back seat — and even this rule of thumb is honored as much in the breach as in the observance.

A business development consultant included the following in his report to a client who had asked for the pros and cons of pursuing additional commissions from governments at various levels:

> A constantly increasing consideration in this type of work is the insistence of the governmental unit client that any design firm from outside the particular unit's boundaries associate with one or more local firms in order to receive consideration for a job. Such a condition is understandable, if not always desirable from a professional standpoint, and, like it or not, it is an ever-growing fact of life.

FORCED ASSOCIATIONS

> One can and should have valid objections to associations where government officials insist on naming the local associates or consultants to be used — but this has been encountered only rarely up to now. Such shotgun marriages may well be met much more frequently in the future, depending upon a number of factors, of which the condition of the economy at the time is probably paramount. In each case the design professional obviously will have the right of choice as to whether or not he wants to enter into such an arrangement.

> At the federal government level, where most of the future design and construction work will at least be sponsored and/or financed, with greater or lesser degrees of local and state participation, insistence on the use of local architectural and engineering associates and consultants is already increasing. In recent months your firm has entered into many more associations and joint ventures than was the case a few years ago, and usually for just these practical considerations.

> Large-scale planning projects appear to be on the upswing. A certain number of such projects are now sponsored by private clients, with government funds almost always involved in one way or another. The expectation is that more and more projects of this type, due to their size, cost and complexity, will be funded totally by government at one or more levels.

ASSOCIATION CASE HISTORIES

Two case histories of theoretical associations should serve to illustrate some of the points we have been discussing.

Case 1

A local high school board has decided to build three vocational technical high schools and a new football stadium to serve all of

the high schools in the district. The board members have concluded that their patrons would be best served by having one
architectural firm act as coordinator-consultant for all four projects. There are several good-to-excellent smaller design firms
located within the boundaries of the consolidated school district,
but none is large enough to take on the whole package. There
are perhaps a dozen large firms in the country with extensive experience in designing vocational tech schools—and many of these
same offices have also designed stadiums, arenas, and coliseums
for a variety of clients and in a variety of locations.

Preliminary cost estimates for the four structures have run as
high as $20 million. The local firm of John Doe & Associates,
having a couple of good contacts on the school board, decides to
pursue the project. Doe & Associates is an eight-man office with
six small completed school projects in its portfolio. Doe's two principals discuss possible associates of national stature and practice
and settle on the Richard Roe Partnership, which has its main office
in New York City. John Doe places a call to Richard Roe.

Let us digress a moment, while the operator is getting Mr. Roe
on the telephone. There is a kind of one-upmanship in putting
together an association, in that the originating party is tacitly
awarded a slight edge over the invitee or invitees. This has no
special advantage in the earlier stages of discussion and negotiations before the job is awarded, but the assembler of the package—
particularly if he is from the smallest of the association participants
—may assume certain prerogatives in the final contract negotiations, often gaining a slightly higher fee or other consideration for
his organizing efforts. As an example of what this could mean,
a difference of as little as one percent in one office's share of the
total fee for a $20 million project could be as much as $12,000—a
worthwhile difference to an eight-man office which might not earn
more than $150,000 in total annual fees.

John Doe is now connected with Richard Roe. In their discussion it develops that Roe has not heard about the vocational tech
school and stadium project and therefore has not been approached
by anyone else about associating for it. If Roe is not familiar with
Doe's firm he will usually ask for a day or two to think it over,
which Doe is glad to grant.

Another unwritten rule of the game of association prohibits Roe
from now going after the job on his own, if he decides against

associating with John Doe & Associates. At least this is an unwritten rule of most professionals involved in business development for their firms. If Roe turns Doe down and is approached later by another local firm about associating for the same project, this, in effect, makes it a new ball game for Roe.

The "day or two" to think over Doe's inquiry therefore is not used to try to work out another association or undercut Doe's position in any way. Rather, Roe will be trying to learn more about the project, assessing its chances of going ahead and, if Doe & Associates is not known to him, getting a fairly complete rundown on the firm. A Dun & Bradstreet commercial report, on a priority basis, would be a minimum check; Roe would probably call some of his friends in the profession for a reading on Doe and his practice. Roe's bank can run a quick credit check through Doe's local bank.

As one who has been involved in putting together many associations, joint ventures, and consortia—both through inviting and being invited—I would make a suggestion to the smaller firm trying to work out an association for the first time. When you call a national firm do not, as some do, try to be cagey or cute about the location and scope of the project, identification of the potential client, or any other significant details about the local situation which an outsider should know in making a reasoned decision on entering into an association with your firm. The chances are that he can find out such details without your help, but remember that you may be working closely with him and his staff for perhaps three years, if your venture is awarded the commission. Be candid and open in every contact—and you will be treated the same way. To act otherwise establishes you as a "bush leaguer" to larger firms.

Assuming that Roe is satisfied with Doe as a potential associate, he calls back to tell Doe he is interested. They agree to exchange brochures, Form 251s and any other material which would be helpful in learning more about each other's practice, staff, and experience. It now falls to Doe to make the initial contacts with his friends on the school board. He may or may not tell them of his association with Roe at this time; his primary interest at this point is to get on the board's list of firms to be invited in for interview. At some point—the earlier the better—Doe will visit Roe or vice versa, depending on individual travel schedules and commitments. Decisions need to be made on the preparation of a joint brochure and other graphics and how the formal presentation will be handled. Some associations make up a special title slide, for example, listing all of the firms in the association, as a way of unifying their individual projects which will follow. A joint brochure is almost always worth the time and effort required to produce it. In some cases it is a requirement of the Request for Proposal.

WORK AND FEE DIVISION

Some discussion probably will be devoted to the eventual division of work and fees. If this is Doe's first association with a large firm he will be much more concerned than Roe about how and where the work will be done and the division of the fee. Roe knows there is not much point in getting into a lot of details on these subjects until they both know more about each other, the scope of the job, its program and requirements, and the client's possible desires as to where and how the design and production will be done. Even though Doe invited this particular carpetbagger in, he may well have some suspicions that the larger, better-known Roe may try to cut him down—or even out—in the clinches. Such concern is rarely warranted, but I have seen baseless and needless worries destroy well-founded associations even before the participants got to the interview.

Doe and Roe probably will discover that a consultant or two should be added to their team. As we discussed in the chapter on Presentations, the consultants ordinarily would not be present for the interview, but Doe and Roe should be prepared with the names of several reputable consulting firms to recommend to the selection committee of the school board. Members of the committee may have some suggestions of their own, as well.

The association of Doe and Roe appears to be moving along on a fairly even keel. The remainder of their joint efforts should parallel those described in earlier chapters, including intelligence gathering, bridge building, and the presentation itself. The differences are (1) there are more hands to share the work in an association and (2) someone should be designated coordinator, to avoid duplication of effort and to achieve the maximum effectiveness and input in the time remaining to them before the interview.

CASE 2

Our first example had Doe, a small local firm, calling in Roe, a large national firm, on an association. The procedure operates in reverse just about as frequently, i.e., a large national firm hears about a project through one of its many contacts and, realizing that it has no chance of getting the commission on its own, approaches a locally based office about an association. Because the large firm has access to many information services, both formal and informal, it may be assumed that its selection of a potential local associate is based on rather careful research.

WHAT TO LOOK FOR

As the representative of Morton, Salt & Cellar, a large Los Angeles-based firm of architects, engineers, and planners, you are particularly interested in several points about Jones, Smith & Associates, a six-man office in Squeegee City, Oklahoma. (Squeegee City is about to hire a designer for its proposed $17 million civic center.)

1. What related projects has JS&A completed in the last five years? (Due to JS&A's size, almost any job involving people handling and movement would be considered to be related.)

2. Who are their prime contacts at city hall and in the ruling political party in Squeegee City?

3. What other city work have they done? (A multi-edged question; if they have done none it might be because they have not pursued city work, in which case it could be their "turn." By the same token, it might be because no one in city hall believes them to be capable. On the other hand, if JS&A has a long record of city projects, the selection committee and political leaders could feel that the civic center should go to someone else, under the "pass-the-work-around" philosophy. These are all questions requiring answers.)

4. What opinion do their fellow professionals have of JS&A—of the principals, the firm's design philosophy, and its general output and dependability?

5. What do some of JS&A's past and current clients think of the firm and its staff?

6. Is JS&A currently involved in litigation with any of its clients? If so, what is the nature of the litigation and could it affect your chances of getting the civic center project?

7. Is JS&A financially sound? How about the health of the principals? It is frustrating, at best, to lose out on a potential commission because your local associate is forced into bankruptcy two

weeks before the interview or the managing partner passes away suddenly a day or two after the interview. I have seen both events happen in associations and, in both cases, what appeared to be a sure job suddenly evaporated.

These are some of the more important areas you should investigate thoroughly before finalizing an association.

FINDING ASSOCIATES

Assuming that you are familiar with the advantages of entering into an association to pursue a specific job, how do you know which firms to talk to in putting together a joint venture with the desired combination of strengths and local or regional contacts? If yours is a smaller firm you would normally be looking for a larger, multi-discipline office with significant experience in the project type and a past history of successful associations. This will not be much of a problem if you regularly follow the news sections in architectural publications such as *Architectural Record, Forum*, and *Progressive Architecture;* you should have a pretty accurate picture of which firms have been the most active in the project type in recent months. If none come immediately to mind a quick review of back issues of the magazines for the last six months or so should give you some ideas. You might also make it a point to talk with some of the salesmen who call regularly on your office. They may have some good suggestions. The listing of architectural firms by geographical location of their offices in the back of the *American Architects Directory* (see Chapter 4) might give you some leads.

Engineering News-Record's annual listing of the 500 largest design firms can be helpful if your primary interest is in locating a large office in a certain area. The problem with this list in its published form is that it can only be used effectively to locate large offices. Firms are shown in descending rank of total billings for the year rather than by location. For anyone who wants to go to the trouble, *EN-R's* listings can be the basis of an invaluable card file list for larger offices to use in locating potential associates for major projects. Get copies of all of the lists from the past six or seven years. Using the oldest list, have your secretary type a file card for each of the 100 or 200 largest firms. It might save some time if you had the lists photostated up to twice or three times the original size and then cut out and paste on a separate 3 X 5 inch card each listing, including the information in the right-hand columns. You will need two complete sets of cards—one for an alphabetical file and one for a geographic breakdown—so you can order two copies of the photostats or simply copy them twice. When you have the first card sets complete, put them into alphabetical and geographic order, the latter by state and city. Then, add each succeeding year's listing to your established files. Be sure to include the relative annual ranking by billings. You will have to add a few new cards for each year because offices inevitably move up and down in the ratings. By the time you finish this exercise you will have the names of several hundred easily identifiable potential associates, together with a breakdown of their in-house services and disciplines. Admittedly, such a list will be of much greater use to a large national firm seeking to locate an associate in a specific city or area than to a smaller office seeking an experienced, larger firm.

Still another source of potential associates' names is the registration lists of attendees at national and regional conferences, such as the annual sessions of federal programs cosponsored by the AIA, CEC, and NSPE. The firms which send representatives to such meetings have tacitly indicated that their interests extend beyond the city limits of their firm's location.

"PERMANENT" ASSOCIATIONS

While most associations are entered into for the pursuit of a specific, single project, some associations are established on a long-term or "permanent" basis. These are usually associations set up to go after a specific project type, such as airports, or to work in a certain area, such as Southeast Asia. In a permanent airport joint venture you might find an architectual firm, an engineering office, a systems group, a research organization, and a specialist in the design of navigation aids. The members of this association are free to pursue any other building or project type on their own or in other associations, but are expected to return to the fold whenever a likely airport project turns up. Amer-Asia Consultants, Inc., is an example of a loose association put together to seek jobs in certain countries in Southeast Asia. We will have more to say about Amer-Asia in the next chapter.

The so-called permanent associations are viewed by some professionals as running counter to the main purpose of an association, which is to organize the best possible team of experts and specialists for the client's needs. Clearly, no job is just like any other in all respects (even the ubiquitous Quonset huts had to be adapted to the site), and there are some airport projects, for example, which require a much higher degree of attention to the runway design than to the terminal or high speed refueling facilities. Other airport jobs may require an inordinate amount of front end work before a pencil is ever taken to paper in design.

Proponents of the permanent association point to the advantage of team members gaining a greater degree of coordination with each job—much as a winning football club must practice by the hour. The input from each firm in the association can be varied to fit the demands of the job, so that individual clients are served according to their needs and the program. My own opinion is that permanent associations are not very effective and my personal preference is to assemble a team to meet the requirements set out in the Request for Proposal.

REQUEST FOR PROPOSAL (RFP)

Depending upon their author and the subject, RFPs can range from very helpful to extremely confusing to their recipients. There is no "best" or "accepted" format for an RFP, but it should contain at least enough information about the proposed job and details for preparing and submitting the proposal so that any interested office can make an intelligent proposal. The following RFP, for a new courts building in Montgomery County, Maryland, is an example of a good request letter:

Gentlemen:
 Montgomery County is soliciting written proposals from architectural firms for the design and supervision of construction of a County Court House to be located in Rockville, Md.
 It is planned to construct the Court House on property owned, or to be acquired, by the County in the Urban Renewal area of downtown Rockville, Md. (Site No. 1, p. 23, and Alternate A, p. 36, of *Building Program Report by Leo A. Daly Co. on the Physical Environment for the Montgomery Country Government Center*).
 Montgomery County will pay a total fee of five and seven-tenths percent (5.7%) of the cost of the project (estimated to be $10,000,000).

Initially, a contract will be awarded for 80% of the fee for work up to and including the award of a construction contract. After the award of a construction contract, the County will contract with the architectural firm for the supervision of construction and the completion phase of the project. Payments and reimbursements will be made substantially as provided for in *Special Conditions of Agreement Between County & Architect.*

To be eligible for consideration, a firm must have:

1. A home office or branch office in the Metropolitan Washington Area.
2. Evidence of continuity of practice.
3. Demonstrated design ability as evidenced by awards and photographs.
4. Adequate architectural, engineering and planning manpower resources.
5. A project manager of demonstrated capability.
6. A license to practice in the State of Maryland.
7. Adequate financial resources to complete the scope of work required.
8. Submitted a formal proposal as required below.
9. A willingness to sign a contract for services, a blank copy of which is enclosed.

Each proposal submitted should contain:

1. A complete and current U.S. Government Architect-Engineer Questionnaire (GSA Standard Form 251). If a joint venture is proposed, each firm must submit a completed Questionnaire form. The form should highlight completed public or quasi-public projects of the type and size comparable to this project.
2. Evidence of design ability for public or quasi-public projects of the type and size comparable to this project.
3. Complete details of any joint-venture proposed for this project.
4. Name, professional qualifications and work experience of the person to be named as project manager.
5. Maryland Registration Number.
6. A statement of design, construction, and liaison problems envisioned and the methods and procedures proposed to accomplish the tasks of design and construction. This requirement may be supported by graphical exhibits.

Single copies of proposals should be submitted to this office, to be received before 5 P.M., July 17, 1970.

After that date, proposals will be reviewed and analyzed by the County and a selection will be made of not less than five nor more than ten of the most qualified proposals.

These most qualified firms will be notified of their selection and will be requested to make an oral and/or visual presentation of their proposal and capabilities to the County. The proposal must be made by the architect designated as project manager. At this time, each firm must submit evidence of adequate financial resources to complete the scope of work required. Financial information submitted will be held in confidence by the County and will not be made public.

After hearing the proposals of the most qualified firms, the County Manager and others from the County may wish to visit typical buildings that have been designed by and constructed under the supervision of the architectural firm. The expense of any such visits will be borne by the County. However, the architectural firm will be required to provide information and guides at its expense.

After visits are made, the County Manager will select the architectural firm and a contract will be negotiated with the architectural firm selected. Other firms which have submitted proposals will be notified of the County Manager's decision.

Any questions relating to this request for proposal should be directed to this office.

Very truly yours,

Lee O. Falwell, Director

Enclosures: Building Program Report by Leo A. Daly Co.
Standard Form of Agreement Between County & Architect
Special Condition of Agreement Between County &
Architect

Another type of RFP is illustrated by the General Services Administration's Request for Preliminary Proposals for construction management services on a project in Beltsville, Maryland:

General Services Administration
Public Buildings Service
INVITATION NOTICE
CONSTRUCTION MANAGEMENT SERVICES
Consoliated Federal Law Enforcement Training Center
Beltsville, Maryland
GS-03B-10563

July 1, 1971   Project No. 190049
A request for Preliminary Proposals (without prices) for construction management services for the Consolidated Federal Law Enforcement Training Center, Beltsville, Maryland, will be issued by the General Services Administration (GSA) on or by July 21, 1971. Preliminary Proposals (without prices) will be received until August 17, 1971, at 4:45 P.M., local time, in Room 1315, GSA Building, Mailing Address: General Services Administration, Public Buildings Service, Office of Operating Programs, Schedules and Services Branch, 19th and F Streets, N.W., Washington, D.C. 20405. The total estimated construction cost

of the project is approximately $36 million. The Government will consider firms meeting the following minimum requirements:

1. Demonstrated experience and competence in performing construction management services.

2. Shall have acted during the past three years as a Construction Manager on at least one project with construction cost in excess of $18 million.

3. Shall be financially able to provide the services required by the Government.

4. Shall have competence in architectural, civil, mechanical, electrical and structural engineering; construction estimating; cost accounting and control; tenant coordination; project management; contract negotiation and administration; construction superintendence and supervision; and other related fields.

5. Shall have constructed buildings in the general geographic area of this project, or have good recent knowledge of local conditions in the project area, or can retain others with such knowledge.

6. Be able to provide professionally qualified personnel to staff the project.

7. Have a good professional and business reputation, and an on-time performance record.

Preliminary Proposals will be received and evaluated on the basis of the requirements and criteria contained in the Request for Preliminary Proposals. Prospective construction management firms who are interested in the project will be furnished the Request for Preliminary Proposals by GSA upon request.

After the Preliminary Proposals have been received and evaluated by GSA, an Invitation for Price Proposals will be requested from only those firms whose proposals have been determined by GSA as being most favorable. Only price proposals specifically requested by the Government will be considered.

A copy of the Request for Preliminary Proposal may be obtained from the GSA Schedules and Services Branch office cited in the first paragraph of this notice.

GUIDING RFPS

A point to keep in mind about RFPs. If you have been on your toes and made early contact with the potential client (usually a governmental unit), often you will have an opportunity to make an input to the writing of his RFP. This is especially true if you have accumulated a file of RFPs from past projects and can come up with several which are pertinent to his job. This would not be the case for federal work, as a general rule, but you may be able to exert a fair amount of influence in county and city RFPs. This is not to say that the end result will be an exclusionary RFP stacked in your favor—but you may gain some elements in your behalf and the client will certainly have to remember your assistance.

REFERENCES

[1] Bernard Tomson and Norman Coplan, *Architectural & Engineering Law,* 2d ed., Reinhold Publishing Corporation, New York, 1967, p. 55.

[2] Justin Sweet, *Legal Aspects of Architecture, Engineering and the Construction Process,* West Publishing Co., St. Paul, Minn., © 1970. P. 92, sec. 4.10.

[3] Henry Campbell Black, *Black's Law Dictionary,* West Publishing Co., St. Paul, Minn., © 1969. P. 73.

[4] By permission. From *Webster's Third New International Dictionary* © 1971 by G. & C. Merriam Co., Publishers of the Merriam-Webster Dictionaries. P. 132.

[5] Ibid., p. 1219.

[6] Ibid., p. 485.

CHAPTER 12

Overseas Client Acquisition

Upon reading the title of this chapter the majority of readers may well be saying to themselves, "Wait a minute, we have not even begun to tap the potential in our own city (or state or country); why should anyone in his right mind go in search of new headaches and unneeded frustrations by pursuing jobs in foreign countries?"

A valid question—and part of the answer is that overseas work is certainly not for every professional design firm. But there always is a considerable amount of work available in other countries and since many design professionals have at least a latent curiosity about how it would be to design, say, a hotel in Kuwait or a hospital in Portugal, it seems appropriate to include a discussion of overseas client acquisition in a book of this type. By the time most readers reach the end of this chapter, their doubts about doing foreign work will be either at least partially dispelled, or else reinforced to the point where they would never consider it. For the firms who have been active outside the United States for any period of time there may be little information of any significance here. If for no other reason, veterans of overseas work may want to read along to compare their experiences with those of others who have been through the mill.

OVERSEAS WORK PRO AND CON

The United States Department of Commerce, which is one of the primary fountainheads of information on foreign work, recently issued a brief paper on "The Pros and Cons of Exporting." With slight adaptation, the paper covers many of the most-asked questions by design professionals on the subject of overseas work.

Con: We have plenty of domestic business.
 Pro: Foreign work spreads risk in times of recession.
 Pro: A satisfied firm is on the road to extinction.
 Pro: So do your competitors, and many of them work overseas.
Con: Our labor costs are too high to compete with cheap foreign labor.
 Pro: High wages do not necessarily mean high cost per unit.
 Pro: The normal procedure is to work with an overseas associate, thus taking advantage of lower labor costs where they exist.
Con: We do not have personnel experienced in foreign work.
 Pro: You may not need a large, specially trained staff to handle foreign projects. If a market looks promising, employment agencies can locate the right people or an executive search organization can pinpoint key international personnel. The local associate or representative should take care of most problems as they arise.
Con: On foreign projects we have to wait too long for our money.
 Pro: Facilities for financing such work have greatly improved in recent years. Consult your banker and the U.S. Department of Commerce.
Con: We cannot compete with foreign countries.
 Pro: Can you afford not to compete? Markets abroad are expanding faster in some countries than they are at home.
Con: Foreign work is too much trouble and involves too much red tape.
 Pro: Red tape is often a problem at home, too, but it can be dispensed with or eased as one gains familiarity.
 Pro: Specialized firms are ready to cut through the red tape for you. The U.S. government also has many services to ease the red tape or to minimize your problems with it.
Con: Import duties are too high in most countries for us to be competitive.
 Pro: Import duties are not your problem. The client pays them.
 Pro: Tariffs and duties are being reduced by GATT and other treaties and what was impossible a few years ago may now be both possible and desirable.
 Pro: There are many countries in the world, and not all of them have high tariffs.
Con: No foreign clients exist for our services.
 Pro: Have you ever looked? Unfounded negativism is hardly a characteristic of a viable, progressive organization.

Con: We don't have time to explore the possibilities of overseas jobs.
Pro: Someone else may spend *his* time studying the subject for you. Visit the U.S. Department of Commerce.
Con: We tried foreign work and had bad experience in collecting our fees.
Pro: Have you ever had a domestic project or client go sour? New aids and better collection devices are now available.

OFFICIAL ENCOURAGEMENT

In the event the reasons for the bullish attitude of our government on American firms selling goods and services overseas escape any reader, the following statement by President Nixon, issued on March 25, 1970, may help to clear up the matter.

In the best tradition of our free-enterprise system, American business and government together must seek and develop every opportunity to meet the challenge of growing international competition.

Success in this vital effort will have an impact not only in the United States but also overseas. For what we achieve will bring returns both for the progress of our domestic economy and for growing social economic development throughout the world.

Selling abroad today is a complex problem. It takes time and consistent attention. The increased effort must, therefore, begin now. Faster export expansion will advance the healthy growth of American business, strengthen our trade position and improve our balance of payments.

WHERE TO START

An excellent starting point for any investigation of the possibilities of pursuing overseas projects is the booklet *Introductory Guide to Exporting* published by the U.S. Department of Commerce and for sale by the Superintendent of Documents, U.S. Government Printing Office, Washington, D.C. 20402. The price is 35 cents. Chances are that the booklet will be found in the nearest Department of Commerce Field Office. More than forty of these offices blanket the United States, stocking a wide range of Commerce and other official government publications relating to business. Each office also maintains a business reference library containing periodicals, directories, publications, and reports from official and private sources.

Within the Department of Commerce the specific responsibility for stimulating, fostering, and encouraging both domestic and foreign business expansion rests with the Assistant Secretary for Domestic and International Business (DIB). His office supervises the Bureau of International Commerce (BIC), the Bureau of Domestic Commerce (BDC), and the Office of Business Services (OBS). OBS is in charge of the forty-two field offices mentioned above.

The DIB organization chart is a little cumbersome to produce here; for practical purposes it is sufficient to know that BIC has the primary responsibility for assisting the design professional interested in overseas work and for promoting exports in general. The Export Business Relations Division is the initial contact point in the Department of Commerce for general questions on how to get into foreign project work, trade problems, financing, markets, investments, and related subjects.

OFFICE OF INTERNATIONAL BUSINESS ASSISTANCE

Specific questions and problems may be brought to the attention of the Office of International Business Assistance (OIBA), whose responsibility it is to assist on a case-by-case basis, thereby speeding solutions to problems faced by U.S. companies in international trade. On the third floor of the Commerce building in Washington, D.C., the OIBA has established a Foreign Projects Reference Room, where U.S. firms can review a wide range of materials on major foreign projects in the OIBA inventory and on projects being considered for financing by the World Bank, International Finance Corporation, Inter-American Development Bank, Asian Development Bank, and the United Nations Development Program. This source room also contains information about the procurement regulations of these international organizations; forms for registering with U.S. and multinational agencies; and information brochures on the programs of the Export-Import Bank, Overseas Private Investment Corporation (OPIC), and the Agency for International Development (AID). The special "Questionnaire for Consulting Firms," as used by the International Development Association, International Bank for Reconstruction and Development (World Bank), and the International Finance Corporation, is reproduced on the following page. See Figure 12-1.

The Inter-American Development Bank's Professional Services

FORM NO. 296.01   INTERNATIONAL FINANCE   INTERNATIONAL BANK FOR   INTERNATIONAL DEVELOPMENT
     (9-69)   CORPORATION   RECONSTRUCTION AND DEVELOPMENT   ASSOCIATION
                                      1818 H STREET, NORTHWEST
                                      WASHINGTON, D.C. 20433

OFFICE OF THE ENGINEER ADVISER – PROJECTS DEPARTMENT

QUESTIONNAIRE FOR CONSULTING FIRMS

NAME AND ADDRESS OF FIRM	PRINCIPALS OF FIRM (Please give names and titles and indicate with whom contacts should be made.)
Organization Date of Firm:	

APPROXIMATE NUMBER OF STAFF ON PERMANENT ROLLS		ESTIMATED COSTS OF PROJECTS IN U.S. DOLLARS	
Professional	Non-Professional	Value of Work in Hand	Approximate Annual Value
A. Domestic:	A.	A. Reports	A.
B. Abroad:	B.	B. Design	B.
		C. Supervision of Constr.	C.

1.   PRINCIPAL FIELDS OF INTEREST

Engineering	Agriculture	Architecture	Economics	Management	Other
☐	☐	☐	☐	☐	☐

2.   PRINCIPAL FUNCTIONS

Feasibility Studies	Development Planning	Design	Supervision of Construction	Operation and Maintenance	Other
☐	☐	☐	☐	☐	☐

3.   ACTIVITIES FOR WHICH FIRM IS QUALIFIED BY EXPERIENCE

From the list attached, select those activities in which you have a particular interest and for which you feel you are especially qualified. List these by symbols in descending order of priority.

No.	Symbol	EXAMPLES OF EXPERIENCE (attach additional sheets if desired)
1.		
2.		
3.		
4.		
5.		
6.		

CATEGORIES UNDER WHICH CONSULTING FIRM SHOULD BE LISTED

4. CATEGORY	CHECK	CATEGORY	CHECK
Independent Consulting Firm	☐	Contractor with design offices which offer consulting services	☐
Firm of Consultants who also undertake functions performed by contractors	☐	Manufacturer of specialized plant with design offices which offer consulting services	☐
Firm of Consultants associated with or owned by contractors or manufacturers	☐	Commercial operating organization or national-ized organization undertaking work as consultants	☐

5. PROVIDE ON ADDITIONAL SHEETS:

     A. A listing of recent projects for which your firm has been responsible.

     B. Designation of those countries in which the firm has been or is working.

     C. Information about key personnel to include technical and professional backgrounds of education and experience.

6. Enclose any brochures or other information, which you may have available and which will serve to define your capabilities.

Fig. 12.1

Contractor Resume, with special instructions for its completion, is reproduced as Figure 12-2.

In assessing the potential for foreign work, a prime source of inquiry is a firm's past and present clients with operations overseas. There may not be so many of these but that the principals would be aware of all such clients on their books. In order not to pass up any possibilities, however, it is wise to extend the research and intelligence gathering to some of the specialized directories in the field.

GENERAL REFERENCE SOURCES

Probably the most complete one-volume listing is the *Directory of American Firms Operating in Foreign Countries.* This 1,006-page directory is compiled by Juvenal L. Angel, Director of the World Trade Academy, and published by Simon & Schuster, Inc. The seventh edition, published in 1969, sold for $35. It is divided into three sections: Part 1 is an alphabetical listing of American corporations with operations in foreign countries, with the overseas addresses; Part 2 shows the geographic distribution of these corporations by countries; and Part 3 is an index of companies by the product or services offered. Should the directory be unavailable in local bookstores, the Technical Reference Department, Simon & Schuster, Inc., 1 West 39th Street, New York, N.Y. 10018, will be able to help.

If the 1969 edition is the latest one available, then the researcher would be better served by turning to the *Trade Lists of American Firms, Subsidiaries and Affiliates,* published by the Bureau of International Commerce, U.S. Department of Commerce. The Trade Lists, each covering a separate country, are compiled for the Commerce Department by American foreign service officers overseas, under the direction of the Secretary of State.

As of this writing, Trade Lists for some eighty countries were available at $1 each. The number of countries covered by the Commerce Department Trade Lists is gradually shrinking, as the field is taken over by private publishers—usually American Chambers of Commerce located in the country. In England, for example, the *Anglo-American Trade Directory* is published by the American Chamber of Commerce in the United Kingdom, 75 Brook Street, London, W.1, England. There is no Commerce Trade List for

England, as a result.   For current list of Trade Lists available, write the Bureau of International Commerce, Export Information & Services, Department of Commerce, Washington, D.C. 20230.

Trade Lists usually are arranged in two parts—an alphabetical listing of companies, followed by a classified list of products and services.   Information given for each company includes the name of the organization, local address and telephone number, name and address of the American parent company or the name of its U.S. owner, type of business in which the local firm is engaged, and the name and nationality of the local manager.

A good general guide to sources of information on business techniques abroad is *International Business and Foreign Trade Information Sources,* edited by Lora Jeanne Wheeler, librarian of The American Institute for Foreign Trade.   Among the topics for which detailed source information is provided in the book are taxation, financing, marketing, insurance, business customs in specific countries, organizations, schools, and investment conditions in more than twenty countries around the world.   Available from the Gale Research Company, Book Tower, Detroit, Michigan 48226, the guide sells for $14.50.

JOB REFERENCE SOURCES

Sources of specific information on foreign work are not so numerous as those for the United States (see Chapter 4).   One jobletter which should be investigated by any firm with a serious interest in getting itself established overseas is the *Foreign Projects Newsletter,* published biweekly by Richards, Lawrence & Company, P.O. Box 2311, Van Nuys, California 91404.   A one-year subscription costs $150; two years, $270.   For airmail service in North America the cost is an additional $5 a year per subscription. Covering government and private developments, the newsletter lists:

- Projects in the planning stage
- Projects in the procurement stage
- Programs of U.S. industry abroad
- Programs of foreign industry abroad
- Programs of foreign industry in the United States
- Loan decisions that generate dollar credits abroad
- Expansion trends in ECM, COMECON, EFTA, CACM, etc.
- Contract awards

SPECIAL INSTRUCTIONS FOR COMPLETING IDB FORM No. 91-A (Rev. 7-70)

An effort has been made to present Form 91-A in a self-explanatory manner; however, should the format of this resume not be adequate to reflect the full background of a consulting organization, additional pages should be used and further data provided. Past experience has shown that the following items need clarification and/or additional emphasis:

ITEM 1. – *"Field of Major Specialization",* and

ITEM 2. – *"Statement of Fields of Activity in which firm is Specially Qualified"*

> These items provide information to assist the Bank in classifying properly a consulting firm, and determining, *in priority,* the types of projects for which a firm is best qualified. Indicate only major fields of interest in *numerical order* of importance to your firm. If one of your main fields is not included, it may be added at the bottom.

ITEM 4. – *"Representative Summary of Professional Staff Specialties"*

> Indicate the number of individuals available to your firm for each specialty listed. If an individual is a specialist in more than one category, note this under "Remarks", Item 20. Particular care should be taken to list professional personnel as working either "Full Time" or "Part Time" with the firm.

> Also show the number of professionals with knowledge of the languages indicated. The average length of professional experience, in years, should also be clearly indicated for each category.

ITEM 5. – *"Language Spoken by Principals"*

> This item applies only to top level officers or principals of the firm, and not to the professional staff included under Item 4.

ITEM 12. – *"Volume of Business"*

> This item refers only to gross fees received by the firm in each major specialty.

ITEM 17. – *"Workload During Past Five Years"*

> Note that this refers to the total cost of the projects handled by the firm, and *not* to the amount charged as fees. The five year average is intended to reflect the *last* five years of work.

ITEM 19. – *"Permanently Associated Firms"*

> List local or international consulting firms with which an association has been permanently established, showing the names and addresses of these associates.

WJH/bld
July 1970

Fig. 12.2

INTER-AMERICAN DEVELOPMENT BANK (IDB)
PROFESSIONAL SERVICES CONTRACTOR RESUME

DATE:

FIRM NAME

ADDRESS

CABLE ADDRESS

FORMER NAME:

YEAR ESTABLISHED | COUNTRY

1. FIELD OF MAJOR SPECIALIZATION	2. TYPE OF FACILITY, PLANT OR PRODUCT		4. PROFESSIONAL STAFF BY SPECIALITIES (Number of persons)	Number of		Number with Knowledge of				Average Length Experience
				Full Time	Part Time	English	Spanish	Portuguese	French	
Administration	Airports	Petro-chemical	Accounting Specialists							
Agriculture	Agricultural	Petroleum	Agronomists							
Architecture	Bridges	Pipe Lines	Architects							
Audit	Cattle	Ports	Auditors							
Cartography	Cement	Power Plants	Accounting Specialists							
Earth Sciences	Coal	Power Transmission	Architects							
Economics	Cold Storage	and Distribution	Auditors							
Education	Dairy Industry	Pulp & Paper	Cartographers							
Engineering — Chemical	Dams	River & Land	Chemical Engrs.							
Civil	Educational	Improvement	Civil Engrs.							
Electrical	Fertilizer	Roads Highways	Economists							
Foundations	Fisheries	Rubber	Electrical Engrs.							
Hydraulic	Food Processing	Sewage	Financial Analysts							
Industrial	Foundries	Storm & Flood Control	Geologists							
Marine	Glass	Subways	Highway Engrs.							
Mechanical	Hospitals	Telecommunication	Hydraulic Engrs.							
Metallurgical	Housing	Systems	Inspectors							
Mining	Industrial	Textiles	Legal Consultants							
Sanitary	Iron & Steel	Tunnels	Management Specialists							
Structural	Irrigation Systems	Water Supply	Marine Engrs.							
Finance	Mining Installations	Other	Marketing Specialists							
Management	Oil Refineries		Mechanical Engrs.							
Social Sciences			Metallurgical Engrs.							
Telecommunications			Mining Engrs.							
Other			Planners							
			Purchasing Agents							
			Railway Engrs.							
			Sanitary Engrs.							
3. BASIC TYPES OF SERVICE			Social Scientists							
			Soils & Foundations Engrs.							
Administrative Management	Geology	Plant Design	Specifications Writers							
Agrarian Reform	Geophysics	Plant Location	Statisticians							
Agricultural Economics	Industrial Economics	Production Planning	Structural Engrs.							
Agronomy	Landscaping	Public Administration	Systems Analysts							
Appraisals	Land Use	Public Finance	Telecomm. Engrs.							
Auditing	Layout Studies	Regional Development	Other							
Business Administration	Low-cost Housing	Soil Evaluation	5. NUMBER OF		6. ESTIMATED CAPACITY					
Colonization	Machine Design	Specification Writing	PRINCIPALS SPEAKING		(Number of Projects)					
Community Development	Maintenance	Supervision & Inspect.	English		Average Construction		No. Projects			
Cost Estimating	Mgt. Accounting	Project Execution	Spanish		Cost in (US $000)		Firm Can			
Economic Feasibility	Management Studies	Systems Analysis	Portuguese				Handle			
Economic Planning	Market Research	Taxation	French		Up to 100					
Economic Policy	Master Planning	Technical Feasibility	7. SCOPE OF		100 to 500					
Educational Planning	Materials Handling	Tourist Studies	OPERATIONS		500 to 1,000					
Equipment Inspection	National Accounting	Traffic Studies	Domestic		Over 1,000					
Equipment Purchasing	National Budgeting	Transportation	International							
Farm Mechanization	Natural Resources	Economics	Latin America							
Fiscal Policy	Organization	Urban Planning	8. NEAREST REPRESENTATIVE TO WASHINGTON, D. C.							
Foreign Trade	Personnel	Other	Name							
Forestry, Mgt. & Mfg.	Administration		Address							
			Telephone							
			9. FINANCIAL RATING OR BANK REFERENCE							
OM, PRA, F: 91A, 7-70										

Fig. 12.2 (Cont.)

While only a few listings in any one issue will be of interest to design professionals, the newsletter is an interesting and practical way to keep abreast of many types of overseas developments — some of which could lead to eventual design commissions.

The novice will gain some idea of the scope and variety of publications available on international business by scanning the *Checklist of International Business Publications,* issued periodically at no charge by the Bureau of International Commerce, De-

10. OPERATIONAL CHARACTERISTICS		11. TYPE OF ORGANIZATION	
Consulting Firm offering only professional services in several fields.		Individual	
Consulting Firm offering only professional services in a particular field.		Partnership	
Consulting Firm which also undertakes functions performed by Contractors.		Corporation	
Consulting Firm associated with, or owned by, Contractors or Manufacturers.		Joint Venture	
Contractor with technical offices which offer consulting services.		Consortium	
Specialized Manufacturer with technical offices which offer consulting services		Subsidiary of:	
Commercially operating organization undertaking work as Consultants		Name:	
Industry with technical offices which offer consulting services.		Address:	
Government or semi-autonomous government organization which offers consulting services.			

12. STATEMENT OF FIELDS OF ACTIVITY IN WHICH FIRM IS SPECIALLY QUALIFIED (LIST IN ORDER OF PRIORITY INDICATING VOLUME OF BUSINESS IN EACH SPECIALITY DURING THE LAST 5 YEARS)

13. EXPERIENCE IN LATIN AMERICA	Argentina	Barbados	Bolivia	Brazil	Chile	Colombia	Costa Rica	Cuba	Dominican Rep.	Ecuador	El Salvador	Guatemala	Haiti	Honduras	Jamaica	Mexico	Nicaragua	Panama	Paraguay	Peru	Trinidad–Tobago	Uruguay	Venezuela	14. EXPERIENCE WITH INSTITUTIONS SUCH AS		
																								Agency for International Development		
																								Export-Import Bank		
																								United Nations–Special Fund		FAO
																								World Bank (IBRD)	IFC	IDA
Has worked or is working in																								European Investment Bank		
																								EEC Development Fund		
																								Asian Development Bank		
Has offices in																								African Development Bank		
																								Organization of American States		
Has representatives in																								Pan American Health Organization		

15. PARTICIPATION IN IDB PROJECTS (GIVE NAME OF BORROWER, NUMBER OF IDB LOAN, AND PURPOSE OF PARTICIPATION)

16. EMPLOYMENT RECORD (GIVE AVERAGE NUMBER OF PROFESSIONAL AND OF TOTAL STAFF DURING PAST YEARS)

	19		19		19		19		19	
	Profes.	Total	Profes.	Total	Profes.	Total	Profes.	Total	Profes.	Total
Home Office:										
Branch Office:										
Total										

17. WORK LOAD DURING PAST 5 YEARS (ANNUAL TOTAL COST OF PROJECTS UNDERTAKEN IN THOUSANDS OF $)

	19	19	19	19	19
Five Year Average					

18. FACILITIES (SPECIAL EQUIPMENT, LABORATORIES, ETC.)	19. PERMANENTLY ASSOCIATED FIRMS

20. REMARKS	
	SIGNATURE:
	NAME (typed):
Page 2 of 2	TITLE:

Fig. 12.2 (Cont.)

partment of Commerce. A few examples of the listings in the thirty-six-page checklist:

Doing Business with NATO, 1968, National Technical Information Service, 12 pp. Free.

Factors in Overseas Investment, NTIS, free. A brief outline of the factors to be reviewed when considering direct investment abroad.

Foreign Business Practices, 1970, NTIS, free. Basic information on some of the laws and practices which underlie export, licensing, and investment abroad.

Information Sources on Marketing Research, NTIS, 30 pp. Free.

NEEC Report of the Industry Committee on Engineering and Construction Services, 115 pp.   Free.

Market Factors in Yugoslavia, May 1969, Overseas Business Report, 12 pp.   Free.

The *Commerce Business Daily,* cited in Chapter 4, is a continuing source of listings concerning overseas work.   Two of the *Daily's* regular features, "Future Construction Abroad" and "Foreign Opportunities"—both normally found on the last three or four pages—are worth watching.   Periodically the *Daily* carries compilations of United Nations Development Program projects (Pre-investment, Stage 2), World Bank and International Development Association-financed projects, and listings from other international and multi-national agencies. These listings usually begin on page 1 and are carried over to following pages as necessary.

Still other sources of information about overseas projects which will be at least partially financed by U.S. agencies are press releases issued by organizations such as the International Bank for Reconstruction and Development (World Bank) and the International Development Association.   Write to the director of Public Affairs of any agency in which you are interested and ask to be put on their press release mailing list.   The following is an example of the type of information which accompanies a press release from IDA:

April 13, 1972

TECHNICAL DATA

PROJECT: Education
COUNTRY: Central African Republic
TOTAL COST: $5.4 million
IDA FINANCING: $3.9 million, 50 years, 10 years of grace, interest free but a service charge of ¾ of 1% to meet IDA's administrative expenses.
OTHER FINANCING: Government
IMPLEMENTING ORGANIZATION: Ministere d'Education Nationale, Bangui, Central African Republic; Cable Address: MINEDUCATION, Bangui.
PROJECT DESCRIPTION: Construction, furnishing and equipping of four new lower secondary schools and extension of an existing one; provision for additional science and practical subjects facilities in

four lycees; extension of the technical lycee; construction of a new secondary school teacher training college; provision of technical assistance for project implementation and preparation of a long-term educational plan.

PROCUREMENT: Civil works, educational equipment and furniture: subject to international competitive bidding.

CONSULTANTS: To be selected.

ESTIMATED COMPLETION DATE: June 1977

Most of the World Bank and IDA releases go into about the same amount of detail, following essentially the same format.

Worldwide Projects and Industry Planning, a magazine "for multi-national management with engineering, financing, purchasing and construction and operational responsibility," is an excellent controlled circulation publication of Intercontinental Publications, P.O. Box 1256, Stamford, Connecticut 06904. Write the circulation manager for a sample copy and to learn if your firm qualifies to receive WPIP. A regular feature in the magazine, "The Washington Rep," has to be one of the funniest continuing columns on the subject. The serious articles are by qualified writers on such subjects as "The Asian Investment Dilemma," "International Strategic Planning: the New Trends," and "Is There Overkill in the Bank Secrecy Act?"

JOB PEDDLERS

Just as there are any number of legitimate and some not-so-legitimate agents or representatives whose business it is to line up U.S. government work for their clients, agents for foreign jobs abound in major U.S. population centers. Many of them peddle semi-legitimate "leads"; in that they have a cousin in the Ministry of Public Works or a brother in their country's Defense Department, who, for a fee, is anxious to help an architect or engineer land a commission. Unfortunately, the cousin or brother often turns out to be a file clerk or a chauffeur, with no more government connections or clout than the corner news dealer. The fee is not usually asked for in advance of landing the job (be very, very suspicious of any agent who does ask for it first), but the designer-client is asked to cover transportation and living costs incurred during the (usually) prolonged negotiations.

The following example of an agreement with an agent is based on an actual contract between a legitimate agent and a design firm for a major project in Europe.   Dates, names, and places have been changed to protect the identity of the parties involved.

AGREEMENT

THIS AGREEMENT made this 1st day of September, 1972, between FOSTER, OBER, SEES AND TRAID, an Ohio corporation, with offices at Cleveland, Ohio, hereinafter referred to as "FOST," and WORLDWIDE BUSINESS LEADS, INC., with offices in New York, N.Y., hereinafter referred to as "WBL."

WITNESSETH:

WHEREAS, in view of, among other things, the unique abilities and qualifications of WBL, FOST desires to retain the services of WBL for the purposes specified hereinafter, and

WHEREAS, WBL desires to perform such services on behalf of FOST,

NOW, THEREFORE, for the consideration, and subject to the terms, conditions and mutual covenants, hereinafter specified, FOST and WBL hereby agree as follows:

1. The term of this agreement shall extend for the period of six (6) months, beginning as of the day and year first above written and for such additional period, or periods, as to which the parties hereto may mutually agree.

2. In accordance with the terms of this agreement, WBL agrees to use its best efforts to assist in the preparation of a proposal for Architectural and Engineering services relating to the new Madrid Medical Complex on behalf of FOST and its joint-venture partners Norman S. Jones and Associates.   WBL and FOST hereby stipulate that WBL is an independent contractor and not an agent of FOST or its joint-venture partners; that WBL shall perform its obligation hereunder independently and on its own initiative, allocating such time and attention to such obligations as may reasonably be necessary to assure their satisfactory discharge and the accomplishment of the goals contemplated.

3. As reimbursement for the performance by WBL of the services specified in the preceding paragraph, FOST agrees to reimburse WBL for travel and other expenses incurred by WBL incident to the performance of this agreement as follows:

A. *Travel:* WBL shall be reimbursed for actual expenses incurred as a result of travel by its staff, beyond points at least 25 miles distant of WBL's offices in New York, N.Y.; which shall include, where actually incurred, air or surface transportation, meals, lodging and other expenses incurred incident to such travel, such as taxi fares, tips, telephone calls and valet and laundry service.   It is understood that International travel will require prior approval of FOST and that such air travel will be performed at air coach rates when available.

B. *Other:* WBL shall also be reimbursed for such other reasonable and necessary expenses as may actually be incurred and directly relating to the performance of this Agreement and excluding such expenses as may be incurred by WBL as a result of engaging in business generally, such as the expenses of maintaining an office and staff, taxes, insurance and professional and social dues and memberships, etc. WBL shall be reimbursed by FOST for such travel and other expenses, net thirty (30) days, upon the submission of proper invoices.

4. FOST agrees that it will not engage the services of another person or firm during the term of this Agreement, for services to be performed by WBL hereunder.

5. WBL agrees to, and represents and warrants that its employees will, hold all information relative to the work required herein in confidence and trust, and to disclose to no person, firm or corporation, except employees of WBL, any knowledge or information pertaining to FOST proprietary information. WBL agrees to submit for FOST prior approval any writings for publication by it or its employees pertaining to any work performed under this Agreement.

6. All data, designs, drawings, blueprints, tracings, plans, models, layouts, specifications and any and all other memoranda which may be furnished by FOST to WBL or its employees, or which may be produced, prepared or designed in connection with any work performed in connection with this Agreement shall remain, be and become the exclusive property of FOST, and upon the termination or completion of the services hereunder, any and all matters in this paragraph referred to, together with all copies and reprints in the possession, custody or control of WBL or its employees shall, upon request of FOST, be promptly transferred and delivered to FOST, and WBL and its employees shall thereafter make no further use or utilization, either directly or indirectly, of any such data, designs, drawings, blueprints, tracings, plans, models, layouts, specifications and other memoranda, or any information derived therefrom, without the prior written consent of FOST.

WBL agrees to take all reasonable precautions to assure that the work carried on hereunder shall be protected against theft, destruction or unauthorized disclosure. WBL further agrees to comply with all applicable State, Federal and local laws, rules and regulations, including security laws and regulations of the United States Government, insofar as said laws, rules and regulations pertain to this Agreement.

7. It is mutually agreed that any of WBL's personnel performing services under this Agreement shall remain employees of WBL subject to its right of direction, control and discipline and shall neither become employees of FOST nor be entitled to any rights, benefits or privileges of FOST's employees. FOST shall, however, acting for and on behalf of WBL, have the right to instruct any such personnel performing services on property of FOST as to their duties, hours of employment, pertinent safety regulations and all other reasonable requirements. WBL personnel shall at all times be subject to the provisions of this Agreement and satisfactory and acceptable to FOST.

8. Nothing contained in this Agreement shall give WBL or its employees any right or authority to create any obligations or responsibility

either express or implied on behalf of or in the name of FOST or bind FOST in any manner or thing whatsoever.

9. WBL hereby releases FOST from any and all liability whatsoever arising out of any sickness or injury to WBL employees or others under it, or from any loss or damage to its property occurring in the course of performing any task under this Agreement.

10. This Agreement shall be governed by, subject and construed according to the laws of the State of OHIO.

IN WITNESS WHEREOF; the parties hereto have caused this Agreement to be executed as of the day and year first above written.

FOSTER, OBER, SEES AND TRAID

By _____

WORLDWIDE BUSINESS LEADS

By _____

It should be apparent that this was a businesslike arrangement from the very beginning. In this case the agent asked for a finder's fee of 3 percent of the project's construction costs.

Any such arrangement should be approached just as it would be if it were a project in the next state. The prospect should be qualified (Is it a real job?) through any and all means open to the design professional. Admittedly, it is a more difficult matter than checking out a rumor about a new IBM plant. Dun & Bradstreet reports may be obtained on foreign corporations and some individuals. The fee is higher than for a domestic D&B report. The U.S. Department of Commerce often can obtain reports from commercial attaches stationed in our embassies and consulates abroad. There is a small fee for this service. Your bank may have a branch in the country; certainly it will have correspondents—and some information may be available from such sources. If the matter is unusually complicated it might be desirable to have an attorney on the scene check it out. Several U.S. law firms have overseas branches; attorneys should be able to put clients in touch with the appropriate contact abroad.

In some cases the finder's fee is added to the design fee as a percentage of the construction costs. If the agent (finder) expects, say, 3 percent for his services and the A-E fee is 5 percent on a $60 million project, the finder will realize almost $2 million, while the A-E gets $3 million for his fee. This fee relationship has been known to upset a designer inordinately—who conveniently forgets that (1) he probably would not have had any chance at the commission without the agent's assistance and contacts and (2) at no time was he coerced into signing or agreeing to any part of the financial arrangements, i.e., he could have backed out early in the game if his financial and professional ethics were really compromised.

WHO'S WORKING WHERE?

Who is doing overseas design work and where?   In May of each year *Engineering News-Record* reports on the activities of the 500 top design firms for the previous year.   In May 1971, 280 (63 percent) designers reported they do work abroad.   In 1970 more than 160 of these firms—two out of three—actually billed for foreign projects.   *EN-R's* 1972 compilation showed that almost half the leading design offices had billings for work outside the United States in 1971.   Foreign billings of the leading architectural and engineering firms in 1971 came to $195 million, a 15 percent increase over 1971 and a 25 percent jump over the 1969 totals.

Europe attracts most of the U.S. architectural and engineering firms, with France leading the continent in the number of ongoing projects.   South America is the second most popular continent, followed by Asia.   Canada, Venezuela, and Indonesia are apparently very attractive to U.S. design firms.

The 1972 *Engineering News-Record* report on the 400 largest U.S. contractors pointed out that 1971 work abroad increased 22.5 percent over the previous year's $4 billion total.   Of the 400 firms covered, seventy-six were working in 105 countries.   Brown & Root, Inc., of Houston, was the top globe-trotting builder, with work in thirty-three foreign countries.   Of its $1.5 billion in 1971 contracts, over $327 million (22 percent) was outside the United States.

GOVERNMENT INVOLVEMENT

Since much of the work overseas in which these firms are engaged is financed or sponsored by one or more agencies of the U.S. government, let us turn our attention to this type of activity.   Dr. Julian Kulski, AIA, AIP, writing several years ago in the *AIA Journal*, had this comment:

> The United States Agency for International Development, which followed a policy of non-involvement in architecture or architectural problems during its long and extensive foreign aid program, recently has shown

signs of marked improvement in this area. For many years, United States funds supported the building of schools, colleges and other structures throughout the world with no, or very little, architectural management and supervision. Instead of contributing to the economic welfare of the recipient country, this construction has often actually held back national development by encouraging poor standards of construction and architectural practice and benefiting a few unscrupulous individuals rather than assisting in total economic progress.[1]

Dr. Kulski referred to U.S. financial support of all types of projects around the world through AID. Another major source of foreign project funding is the United Nations Development Program (UNDP). The U.S. share of all UN activities has always been a relatively heavy one. Through the thirteenth session of UNDP, in January 1972, more than 1,500 major pre-investment projects had been approved, with total announced costs of almost $3.7 billion. UNDP contributions were $1.5 billion, or 40 percent, with the recipient countries responsible for the balance.

The UNDP program is carried out by the United Nations and eleven other international executing agencies and organizations. These are:

    International Labor Organization—Geneva
    Food and Agriculture Organization—Rome
    United Nations Educational, Scientific and Cultural Organization—
        Paris
    World Health Organization—Geneva
    International Bank for Reconstruction and Development—Washington,
        D.C.
    International Telecommunications Union—Geneva
    World Meteorological Organization—Geneva
    International Atomic Energy Development—Vienna
    Inter-American Development Bank—Washington, D.C.
    United Nations Industrial Development Organization—Vienna
    United Nations Conference on Trade and Development—Geneva

Getting even basic information about UNDP and other UN projects in time to pursue any of them can be a frustrating experience. The Bureau of Domestic Commerce in the Department of Commerce issues a list of UNDP pre-investment projects, following their approval at the January and June meetings each year. UNDP summaries giving some background information about each project are available upon written request from the Office of Business Services, Trade Opportunities Section, BDC-584, U.S. Department of Commerce, Washington, D.C. 20230. The complete project number, as "DP/SF/R.13/Add. 2," must be included, along with a preaddressed label, in all requests for summaries. It is, frankly, a good thing that U.S. firms overseas do not have to depend on United Nations jobs — despite that agency's dependence on U.S. financing.

Information about the organization and purpose of such international agencies as the Asian Development Bank, the Inter-American Development Bank, the Overseas Private Investment Corporation, the Export-Import Bank of the U.S., and the Agency for International Development may be found in the *U.S. Government Organization Manual*, cited as a general reference source in Chapter 4.

OFFBEAT VENTURES

The Office of Foreign Building Operations (OFBO), Department of State, is responsible for the construction and maintenance of U.S. consulates and embassies and certain other overseas government facilities. The OFBO, more so perhaps than most other Washington departments and agencies, has a continuing problem in getting its projects funded by Congress. (A new office building for the House of Representatives is much more practical and visible, for instance, than a functional, impressive embassy building in some far-off country.)

Public Law 480 funds are composed of blocked currency owed to the United States by certain countries and by law must be spent in the country of origin. American officials, such as congressmen, may draw on these funds for expenses and incidentals when traveling in P.L. 480 countries. In late 1972 the P.L. 480 country list included

Burma	Pakistan
Guinea	Poland
India	Tunisia
Israel	U.A.R.
Morocco	Yugoslavia
Nepal	

Now, what has one (OFBO) to do with the other (P.L. 480)? P.L. 480 funds may be tapped by the OFBO to pay for the design and construction of consulates and embassies in the P.L. 480 countries, without having to obtain congressional approval of the expenditures.

Like most ventures today, the procedure is a little complicated — but workable. As an example, let us assume that an American design firm has a use for Yugoslav dinars—perhaps for the purchase of a tourist hotel in Dubrovnick as a short- to medium-term investment. If the OFBO has a consulate project in Yugoslavia at the same time, the design firm may be given the commission with the total fee payable in dinars from blocked P.L. 480 funds. Since there are restrictions on taking dinars out of Yugoslavia and the dinar is not a very convertible currency at best, it is understood from the beginning that the professional's fee will remain in Yugoslavia.

The fee is used to buy into the resort hotel, as above. With any luck, Hilton or Sheraton or one of the airlines will decide it needs a hotel on the Adriatic and the designer can sell the hotel for dollars if he wants to. He then has his fee back in the United States in spendable dollars and the OFBO has its new consulate in Ljubljana or wherever.

This type of transaction is familiar to Western businessmen who have dealt with East European countries and other nations with nonconvertible currencies. It is a barter system, in effect, where machine tools from Cleveland are exchanged for Hungarian canned plums, which are in turn traded for a quantity of cameras made in East Germany. The cameras then are exchanged in Paris for a shipment of men's shirts. The shirts are sold in London for English pounds and once again the salesman is back in convertible cash and with a profit overall, if he has been astute and lucky in his dealings.

By working in conjunction with other international agencies such as the Overseas Private Investment Corporation (OPIC), a firm conceivably may get into an investment along the lines of the Dubrovnick hotel with little or no outlay of personal or company funds. OPIC, the successor to the Office of Private Resources of the Agency for International Development, participates in project financing through loans from its own resources and by guaranteeing loans from private U.S. lending agencies. The corporation also finances pre-investment surveys (feasibility studies) and insures certain overseas investments against such risks as expropriation, war damage, and insurrection.

THE OVERSEAS CLIENT

A feel for working in some areas overseas, especially Asia, may be obtained from the pooled experience of practitioners who have spent many years in countries such as India, Pakistan, Korea, and Hong Kong. Their experiences may or may not parallel those of others who have worked in the same areas.

Although buildings have been erected for centuries in many foreign countries, the U.S. designer is apt to find little sophistication on the part of private clients overseas. A client from Saudi Arabia noticed a detail in an airport restaurant in Washington, D.C., where he was having lunch with his American architect — and immediately wanted the detail incorporated in his building, which was neither a restaurant nor an airport terminal. Clients often change their minds seemingly on just such a whim — but do not understand why they should pay for the changes. This attitude may relate to the fact that monumental works in many parts of the world required literally centuries to complete. Day-to-day changes obviously meant little to the planner and builder. In projects of major size, where the total design fee may be proportionately higher, clients can be even more certain they should be able to make changes at no additional cost. Since construction costs in some countries run up to 50 percent less than for the equivalent building in the United States, a fee as a percentage of costs can be much lower than one would expect.

One small irritation in many jobs abroad is the client with a large staff of employees who wants his own staff utilized by the designer — regardless of training or experience.

THE OVERSEAS ASSOCIATE

Unions are no problem in most Asian countries. Manpower is cheap, while materials are expensive. In general, graft and bribery are a way of life. In some countries the sketch drawings must be translated into local languages, such as Arabic. This is a responsibility of the local associate, who may then try to parlay this "experience" into getting the position of job superintendent. Because of the generally lower fees, the American firm usually depends on the local associate to do more of the work, up through construction documents. Everything must still be carefully checked, both through the designer's representative on the scene and by the project architect back in the United States. This factor, coupled with the undependability of mail delivery, can play havoc with schedules.

The local associate often tries for quantity rather than quality in the drawings, on the theory that the more sheets, the better the job. On one job a foreign associate drew 500 sheets for the project. The job had to be redone later by the American lead architect in his New York office and the domestic total was less than 200 sheets. Where a U.S. firm will normally do few full-size drawings, e.g., a molding detail for the artisan, architects overseas are inclined to work much more at full scale. Even door details sometimes come through full size. Structural drawings are much more detailed, due to the fact that shop drawings often are not done. The full length of a beam, showing placement of re-bars, etc., may be detailed. Mechanical and electrical drawings from foreign associates are often inferior, possibly because these trades are not as well understood; less training, fewer reference sources, and less need for sophisticated layouts are all plausible reasons for the problem.

Specifications as written by local associates in some countries may be quite different in their approach in that the associate tries to retain control over many seemingly insignificant details. It should also be recognized that local associates will often ask for a remarkably low fee, believing that they can come back for extras from the base architect as changes are made. As an example, the foreign associate stipulated in one case that he wanted full control of the execution of the contract. A contract provision required shipping a truck from Beirut to the job site, and the associate insisted on having his representative accompany the truck to insure its delivery on time to the proper site. He naturally expected to be paid for this extra service. If the truck had become involved in a traffic accident, requiring a payment to the local police to take care of the matter, the bribe would be charged back against the base architect along with the salary and living expenses of the associate's representative. Such rigid controls are also applied to the transport and delivery of building supplies.

BARGAIN—BUT WITH HONOR

Negotiating (bargaining) is part of doing business in Asia, Africa, and a few other areas; everyone expects to haggle over prices of everything from the price of a bicycle tire to the design of a luxury hotel.

It is sometimes difficult to get the fact across that American architectural fees are established on a professional basis and associates and clients are expected to abide by their fee agreements. This is one point about which the American designer must be very strict — or else be prepared for trouble all through the execution of the work. A majority of architects in Asia tend to drift easily from established professional fee schedules, where they exist. Generally, local owners expect to pay a higher fee to an American architect, but only after going through the usual bargaining process. It is almost always preferable in most overseas areas for the American designer to establish a flat fee for his work, rather than settling for a percentage of the construction cost fee. This avoids several problems, including reduced fees because of lower construction costs.

In any foreign work the design professional must scrupulously avoid giving any indication that he is interested in or open to bribery. It is far better, working on a flat-fee basis, to stay out of the cost picture altogether. Let the owner handle selections from "or equal" bidders, unless he asks for professional advice on quality or durability. Even this can be dangerous, if the losing supplier is the owner's brother or nephew.

A local attorney should always be used to draw up working contracts and to advise on differences and nuances of local laws and business customs. Both the contract and advice are reviewed with one's U.S. attorney, of course.

Architects and engineers in foreign countries are inclined to look up to their American associates. Architects from Asian countries who have studied in the United States enjoy high esteem among their countrymen and fellow professionals. Most Asians would prefer to take their professional studies in the United States, and if they were unable to, regard an association with an American designer as a desirable form of postgraduate training.

One of the most important things to remember when practicing in foreign countries is that the client must feel his architect is an honorable man. This necessity to conduct a practice on a pedestal has given some American practitioners a type of inferiority complex: they fear that they will be unable to live up to the client's expectations because they are not as qualified as he assumes them to be. As important as honor and ability is tact. Losing one's temper with subordinates can be as much of a faux pas in Eastern countries as belching in public would be in Philadelphia. Learn local mores and customs as quickly as possible.

An Asian will try to avoid "loss of face" in any situation; he feels it is personally degrading to say "I don't understand." He may also consider it impolite (likewise causing loss of face) to say to his American associate, "You have not explained the point clearly enough." The tendency, therefore, is for him to say "Yes" or "OK," rather than to ask for a further explanation or clarification, no matter how important the point may be. Most of us do the same thing occasionally for a different reason. So be as clear as possible when giving instructions or answering questions. Review complicated or difficult points when talking to a client, a local associate, or contractors until it is certain that everything has been understood.

U.S. designers may find their foreign associate not as well trained or as sophisticated as a domestic associate might be, but the overseas man will be anxious to learn and will come to the project with an open, eager mind. Many Americans come to learn that these conditions can make for a more satisfactory working and professional relationship than working with U.S. associates who may bring much more experience and some reputation to the job — along with a closed mind.

LOCAL CUSTOMS AND MATERIALS

Local customs must also be taken into account when designing in foreign areas. The women in most Asian countries hang their wash outside to dry. When designing urban apartments the architect must allow room for bamboo poles to extend from the windows in all directions — especially over courtyards — for the drying of clothes.

Locally supplied materials and the quality of native craftsmanship are almost always excellent. If the U.S. designer is able to travel through the country before beginning design, he and the project may benefit by finding native techniques and items which can be incorporated into the building. This often works to better advantage than if equivalent materials were imported from the West. At any rate, the U.S. practitioner should always investigate materials and methods in a foreign country. Mosaic tile and the labor to install it are extremely cheap in Hong Kong, so one uses the 1-inch tile on fire stairs, for example. Concrete stairs with steel nosing could cost up to three times as much as the tile installation.

Above all, remember that bid documents and design detailing usually must be done to accommodate suppliers and contractors from all over the world.

VIEW FROM INDIA

In its inaugural issue in January 1970, the Bombay-published *Building Practice* magazine carried an interview with S. K. Nadkarni, Bombay architect and engineer and the president of Practicing Engineers, Architects and Town Planners Association (PEATA). Mr. Nadkarni's remarks illustrate some aspects of practice in his country.

> I would say, with all humility, that every conscientious work well done acts as a sort of advertisement and is bound to attract further business. There are many instances when one client introduced me to another client. I would advise newcomers [to the profession] to give full service to their clients, even if they were to get lower fees. A satisfied client helps the professional. The architect, however, should ask for his fees in proper installments, as the clients often let the architect down.
>
> . . . although 1959 to 1966 was a period of general boom in building activity, architects had to put in a lot of labor filling in innumerable forms which the controllers required. Obtaining municipal approval for plans has become very irksome these days. There are a number of forms to be filled in and many times the Development Rules are amended by circulars. By the time we prepare the plans in conformity with the Department Rules and submit them to the Municipal Corporation, the Rules will have been further changed, requiring amendments to the plans. This happens very often. The general moral standard of the public has gone down. Values have changed. The contractors do not carry out their work properly and not a few architects are indifferent. Owners do not pay the architects properly. . . . The whole thing is a vicious circle. Human values have gone down and a tendency to make money somehow is increasing and ultimately the work suffers. I am at a loss to say how the situation can be improved.[2]

CODES

Through foreign—mostly European—influences, much of Asia is heavily coded and zoned. Where codes were not in force prior to World War II, as in Japan for example, the pyramiding effect of ever-increasing population density has forced their imposition. Where they are in effect, codes and zoning are strictly enforced as a rule.

OVERSEAS CONSORTIA

Amer-Asia Consultants, Inc., is an interesting experiment in locating and producing foreign work for small- and medium-sized U.S. consultants who wish to perform abroad. Funded by an interest-free loan from the Commerce Department's Joint Export Association Program, the eleven member firms who make up Amer-Asia are trying a consortium approach to overseas jobs. As H. Peter Guttman, chief executive officer, explains, "Amer-Asia is one opportunity the Department of Commerce has offered independent consultants to challenge competition from foreign groups that have enjoyed increasing assistance from their respective governments."

The consortium maintains its head office in Washington, D.C., and a regional office in Hong Kong. Combined professional, technical, and support staff of the associated firms is in excess of 1,500.

Guttman's point about foreign governmental assistance to their own design professionals is well taken. Some European governments now offer financing for turnkey and similar projects at interest rates so low that the client cannot afford to insist on first-class design.

CARNETS

Once an overseas client is signed up and design is underway, the American architect or engineer may find himself in a monumental hassle with customs officials the first time he attempts to take in drawings, material samples, or equipment. Until a few years ago there was not much of an alternative except to pay whatever customs duties were imposed and hope the charges could be recovered from the client. In 1969 the United States entered into a series of international agreements which allow the design professional to avoid delays and extended customs procedures by use of the ATA Carnet System.

A carnet, for those who never had the experience of taking an automobile into Europe in post-World War II days, is a green folder containing customs papers. Ordinarily there is a pair of customs clearance papers for each country to be visited, one for entering, the other for departing the country. These are removed by customs officers at points of entry and departure. A pair of vouchers is also provided for leaving and returning to the country of origin.

In the United States the issuing organization for ATA Carnets is the U.S. Council of the International Chamber of Commerce, 1212 Avenue of the Americas, New York, N.Y. 10036. Practically every country in Western and Eastern Europe except Poland are members of the Carnet Convention. The cost of a carnet runs from $50 to $149, with the exact fee dependent on the value of the goods to be covered. In addition to the fee an exporter needs either a letter of credit from the U.S. Council or a bank guarantee sufficient to cover the sum of duties and excise taxes, plus 10 percent, in the highest tariff country to which the carnet holder will be traveling.

INTERNATIONAL REGISTRATIONS

An obvious way to overcome many of the problems and harassments which face a U.S. designer practicing overseas would be some type of International Council of Architectural Registration Boards (ICARB). There seems to be little hope of that coming about in the foreseeable future but a chink or two has appeared in the armor of protectionism worn by foreign design professionals. Reciprocity in architectural registration is now offered by Great Britain and a few American architects have passed the examination. As members of the Royal Institute of British Architects (RIBA), they are licensed to practice in England and other Empire countries. A principal of a New York City firm sees his British registration as an opening wedge for client acquisition in such former colonies and English-oriented areas as Australia, Near-Eastern nations, and Canada.

Professional practice in England, as a member of the RIBA, will hold a few surprises and some frustrations for U.S. design professionals. More than 85 percent of British architects belong to the RIBA and are bound by that organization's Code of Professional Conduct. In addition, all persons practicing architecture in England are required by law to be registered with the Architects' Registration Council of the United Kingdom (ARCUK) and are subject to ARCUK's Code of Professional Conduct.

Only in recent years has the RIBA permitted its members to retain public relations counsel. On the subject of business development, the RIBA's Information Office furnishes the following:

1. British architectural firms are prohibited from retaining a person or persons whose primary duties are to discover new projects and then follow up with the potential client on behalf of their firm,

i.e., client acquisition activities, as known in the U.S., are not allowed in Great Britain.

2. RIBA members obtain new work through (a) recomendations, (b) the RIBA Clients Bureau, (c) reference to entries in the RIBA *Directory*, and (d) practice information sheets which the RIBA permits firms to send to local authorities.

3. The RIBA Council is concerned with preventing unfair competition between architects and upholding the principle that a professional man must rely for his advancement on the merit of his work and the personal recommendations of others, and not on self-advertisement.

Regarding business development and public relations, the ARCUK Code states, in summary:

1. An architect may not advertise or solicit for business, nor allow any members of his staff to do so, nor offer remuneration for the introduction of clients.

2. An architect must not solicit publication of articles or illustrations of his work outside the professional press.

Certain interpretations of these sections of the ARCUK Code have been made, including these:

1. In the opinion of ARCUK it is permissible for an architect to engage the services of a public relations consultant to carry out, whether by word or deed, only those activities which the architect himself is permitted to carry out without transgressing the Code of Professional Conduct.

2. Therefore, at the outset of employing a public relations consultant, the architect should hand over to him a print of the Code and ensure that he is fully aware of its Principles and Examples.

3. In his choice of, and instructions to, the public relations consultant, the architect should exercise the utmost circumspection, not only to safeguard the good repute and standing of the profession, but also because the architect would be held answerable for any act committed by the public relations consultant in the course of his employment by the architect, which if committed by the architect himself would constitute a breach of code.

4. When instructing a public relations consultant, the architect's primary consideration and purpose should be to advance the interests of the profession as a whole, and only, secondly, his own legitimate interests.

Rosetta Desbrow, director of the London-based Desbrow Public Relations firm, is one of the few English PR professionals to attempt to work within the strict guidelines set up by the RIBA and ARCUK for publicizing the work of their members. In a memorandum on the subject prepared for the guidance of architects and public relations practitioners, Miss Desbrow pointed out two anomalies in the present ARCUK Code and its interpretations:

1. While articles on general architectural subjects may be written by architects and published under their own names, they may not be illustrated with examples of their own work. This can lead to the curious situation whereas an architect's article is published because he is an acknowledged expert on a particular type of building or environment—or indeed recognized as the leading expert—and the very buildings by which he acquired his expertise and for which he is acknowledged must be left out and perhaps inferior illustrations used.

2. Architects or their public relations advisors may approach . . . a features editor regarding an article, but they cannot approach an architectural or planning correspondent regarding a building which may illustrate a particular point or have some interesting value in itself. They may not approach a writer whom they know to be interested or to ascertain his interest in a building they have designed. But the builder —or even a subcontractor who has played the most minor part in it— may do so.

Readers interested in more information about registration and practice in England might contact the Royal Institute of British Architects, 66 Portland Place, London W1N 4AD, England, or the Architects' Registration Council of the United Kingdom, 73 Hallam Street, London W1N 6EE, England. AIA headquarters in Washington, D.C., will also be able to be of assistance.

When your firm has finally landed an overseas project, and assuming it is a sizable job, one or more members of the firm will no doubt spend from a few months to several years overseas. An overseas move is never simple (take it from one who has made a few) and a helpful booklet is *A Guide to Moving Overseas*, published by United Van Lines, Fenton, Missouri 63026. The guide lists special documents needed for settling in various countries and where to get them. Other topics such as a family's personal needs overseas and establishing banking and credit arrangements are also discussed. The various taxes imposed by U.S. and foreign governments are explained.

Finally, a few DOs and DON'Ts for those working and living abroad:

DO make certain that all points in regard to exactly what the design fee covers are spelled out in detail—both orally and in the written contract.

DO let it be clearly understood from the beginning that all extras will be charged for.

DO be careful with translations. Marlene Dietrich's shoulders were once described as being "white as *Schlag.*" In German, Schlag can mean birdsong, apoplexy, punch, thunderclap, the chiming of a clock or, in street slang, whipped cream, as intended in this case.

DO use tact and politeness in dealing with associates, other professionals, and government officials in foreign countries.

DO expect to be asked many questions about yourself and the United States, some dealing with personal matters such as salaries and costs of various goods and services. Be prepared for lengthy political discussions—and the exercise of patience and diplomacy.

DO make it a point to spend some time in studying the history, customs, and language of the country in which you will be working.

DON'T expect a U.S. attorney to be any more familiar with contracts in, say, Yugoslavia, than you are. Use local lawyers for advice and checking.

DON'T overlook the many services of the U.S. Commerce Department and other government agencies which can be useful in searching out and qualifying foreign work.

DON'T forget the nearest U.S. embassy or consulate in emergencies, or if you need general advice on business or legal matters.

REFERENCES

[1] Julian Eugene Kulski, "The International Architect," *AIA Journal*, July 1968, p. 61.

[2] "Interview with Mr. S. K. Nadkarni," *Building Practice*, Bombay, India, January 1970, pp. 4–5.

CHAPTER 13

A Piece of the Action

Over the years, the entry of design professionals into full partnerships in design and development teams has become known as taking "a piece of the action." Since construction regularly accounts for about 10 percent of the Gross National Product of the United States ($100 billion in 1971), there is a lot of action to take pieces of.

Incidentally, etymologic research on the phrase "a piece of the action" turned out to be singularly unsuccessful. "Action" alone can mean a share of stock, as well as the price fluctuation or trading volume of a security. Another definition for action is the entire process of betting, offering a bet, its acceptance, and determining the winner of the wager. "Piece" is defined in one dictionary[1] as "a share; a financial interest in any business, entertainment or gambling project." But nowhere in the extensive collection of dictionaries of American-English usage and American slang in the Library of Congress could "piece of the action" be found.

Nevertheless, any member of a development team knows exactly what a piece of the action means. That it involves elements of gambling would not be disputed by many who have taken a piece of the action.

WHERE THE ACTION IS

Getting a piece of the action may range from a designer leaving part of his fee on the table in exchange for a small share of ownership in the project to taking on the full responsibilities (and risks) of the developer—including land ownership, project financing, managing the building through design and construction—to ownership and leasing or selling of the completed structure. The ultimate in "action" is a kind of triple-threat consortium, in which the individual or partnership finances, builds, and owns, with all that the three roles imply. Technically, this approach no longer qualifies as a piece of the action; the developer has all the parts and pieces, i.e., the whole action. Fed up with the almost passive role of building cosmetician to which they have been relegated, some of today's professional designers are moving aggressively into a "design-construction-own" posture.

What has happened to the traditionally rather staid, no-nonsense professionals, such as architects and engineers, to push them into the strange and untested territory of real estate entrepreneurism, wherein one assumes the risk for the sake of the profit?

FAULTS OF THE SYSTEM

The point to be made here is that some time ago owners (particularly developers) came to the realization that there was something inherently wrong with a system that required them to wait twelve to twenty-four months while their designer plodded deliberately through concept, preliminaries, schematics, and construction documents. With another two to six months tacked on for bidding (the higher figure for rebidding) and, in some cases, many more months for the architect or engineer to redesign when the bids exceeded the budget, the pressures tended to build on all concerned. All during this period, of course, construction costs steadily increased from 8 to 12 percent a year.

After all these preliminary steps to building were out of the way, the prospective owner was then faced with further delays—strikes, late material deliveries, inability of the contractor to get or keep workmen and equipment. George Kassabaum, a former AIA president, told the Washington Building Congress in June 1969:

> We are part of the only industry that cannot supply enough goods and services to meet the demand. That's why change must come to the construction industry. . . . What does today's process have to

offer? Tradition—but these are fast-changing times and the old way of doing things is being discarded very rapidly in other fields. Competitive bidding does still guarantee the lowest price for any project on any given day, but my own experience is that with prices jumping one percent a month, more and more owners are willing to negotiate to save time, rather than be sure that their price a month or so from now will be the absolute lowest. *What makes me uncomfortable about the conventional way of building is today's building industry's apparent inability to make constructive proposals in the face of specific and urgent problems.*

We commented in an earlier chapter about the increasing importance of governments at all levels as current and future clients. After many years of pretty much taking what they could get in the way of design, costs, and schedules, and usually putting up little or no argument unless an alert legislator raised a question about unconscionable cost overruns, government buyers of construction are also demanding changes in the system.   It is no secret that part of the problem with public buildings lies with the client.   Gurney Breckenfeld, in a widely heralded *Fortune* article of a few years ago, pointed out that nonarchitects often have so much to say about publicly subsidized buildings programs that about all that is left to the architect is a choice about the color of the brick and how to decorate the lobby.

There are any number of horror stories about federal building design and construction. GSA's Public Building Service found that the average $10 million building took sixty-four months to complete when the government was the client, compared with two years for most projects in private industry. New York State buildings used to average seven years from the establishment of need to occupancy, while comparable privately financed buildings went up in two to three years. The Department of Health, Education, and Welfare cites a communicable disease center near Atlanta as a shining example of red tape, confusion, and delay. The center required seven years to complete and, partially because of scientific advances during those eighty-four months, it was obsolete on the day the tenants moved in.   Some $4 million more was spent to bring the building up to date.

THE BUILDING TEAM

Finally, all the dissatisfaction with the established system and the general unrest it engendered among owners and users began to force the building team (as differentiated from the development team) into new ways of designing and building. Fast track, value engineering, construction management, and other esoteric terms began to find their way into the language of the construction industry—of which architects and engineers were and always will be important elements. At the same time, the hallowed design professions were taking their licks from other critics besides owners.

In the *Fortune* article mentioned above, Breckenfeld made some fairly harsh observations about the state of the design art:

> . . . the impact of architects on twentieth-century America has been disappointingly small. The architectural quality of most new building has ranged from mediocre to deplorable. . . . the average house or apartment, store or skyscraper, school or civic structure has more often been a blister on the landscape than a work of grace, charm and good order.[2]

CLIENT COMPLAINTS

A large corporate client recently listed four ways in which he felt architects all too often demonstrate a lack of integrity in their relationships with clients:

1. The partner-in-charge of the project disappears after the contract is signed, never to be seen again during the life of the job.
2. Hiding poor research behind closed specs or meaningless performance specs.
3. Improper coordination of all trades—especially at the interface—and then refusing to accept full responsibility for the lack of coordination when it occurs.
4. Dragging out a time-tested solution they have used a dozen or more times, knowing that it works and has probable adaptability to the project at hand, and then demanding a full creative fee.

The erosion of the professional designer's influence on building has reached the point where perhaps less than 40 percent of all construction dollars is spent on architect-designed structures. Some estimates have it that not more than one building in five involves an architect. "The architects' role in what remains, mostly apartment, commercial and public building, is all too often constricted by fast-buck developers, ignorant but adamant clients, or rigid government procedures and red tape."[3]

RESTRICTIONS ON THE DESIGNER

Buckminster Fuller stated the problem in his inimitable phraseology:

> The client who retains the architect tells him that he has already determined to produce a mortgage-bank prelogisticated building for a specific purpose on a specific site for a specific sum of money. The architect has no original design initiative, for he must design the kind of building the client already has in mind. Usually he must please not only the client, but the client's wife, business partners, lawyers, bankers, real estate agents, and their respective committee of "experts."
>
> By and large the architect has to work with what is available in the industry's architectural and engineering building-component catalogues. Furthermore, he must design within the already preconceived building codes, the zoning laws, and the restrictions set up by various labor unions. The architect also is greatly controlled by what the organized contractors can do, the kinds of tools they have and the price that they bid to do the work. The architect is, finally, just an esthetical, good-taste purchasing agent and ways-and-means detailer—who is practiced in finding and organizing the usable space amongst the columns, pipes, and elevator stacks.[4]

KLING'S TEN COMMANDMENTS

To be sure, a few voices have been crying out from the construction wilderness—some for a surprisingly long time. Almost two decades ago, Vincent Kling went on record in the *AIA Journal* (May 1957), among other places, with his recommendations for expanded services architects should be prepared to offer clients to offset the inroads being made by package builders. In light of the current unhappy state of the profession, Kling's ten commandments for his fellow professionals strike a contemporary note:

> 1. A careful determination of the cost and quality of the job before construction is undertaken. This will thoroughly acquaint the owner with the scope and quality of his project and eliminate the possibility of progressive downgrading through the construction period in order to meet the established budget. The owner is thus protected against buying a "pig in a poke" building.
> 2. Analysis of the relationship of cost to revenue producing capacity of the project before launching the working drawing phase, thereby assuring the economic soundness of the project.
> 3. Quality control over selection of materials and equipment and control over performance of contractors and subcontractors to assure the best result for the owner, who generally is inexperienced in the intricacies of the construction industry.

4. Production of a building which has high resale value by virtue of its flexibility in planning and its superior standard of workmanship and design.

5. Broad service in site selection and use, establishing the suitability of the site for the project and its relationship to the neighboring community. The architect must necessarily broaden the scope of his influence to include considerations of urban and regional planning.

6. Knowledge of advances in building techniques stemming from the broader experience of the architect whose practice involves many different types of buildings and systems of construction. Careful analysis of new products and methods and the courage to employ them when the occasion is right. This progressive approach to building design and construction will prevent the adjustment of a design to a builder's limitations or preferences and, at the same time, bring together the forces which spell progress in architecture.

7. Analysis of the insurance and tax structures as they apply to the design of the project.

8. Coordination of the interior design with the general concept and objectives of the principal architectural design.

9. Preparation of top-quality graphics and public relations material which will complement and support the project.

10. Employment and coordination of consultants in such specialized fields as hospital planning, kitchen planning, and industrial planning, where the best interests of the project will be served. The architect is in the unique position of being able to coordinate these professionals within the overall concept of the project. He alone is qualified to interpret that concept.

In general, it must be pointed out that the architect is the one person on the whole building scene who can serve the client's interests best because he is under no compulsion from manufacturers or contractors and is the only logical person to appraise the merits of the overall project before its design is undertaken.

Kling, in a keynote address to the AIA Central States Regional Conference in October 1971, some 14 years after his proprioceptive analysis of the problems of the body architecture, said:

. . . it is the architect who is charged with the responsibility for designing the total physical environment for people. To do this well he must hold the key position as coordinator and leader. He must be expert in real estate, planning, financing, mechanical, electrical and structural engineering, construction supervision and interior design, as well as performing his unique role of producing architecture of high quality design. To do this he must readily adapt to his role as a developer—a positive force in designing tomorrow's environment.

EFFORTS TO EDUCATE

The designer has increasingly been drawn into the rarefied atmosphere of mortgage banking, property management, syndication, and cap rates—as his clients and others in construction and development have begun to educate him about how much more money is to be made if the building is ready for occupancy in nine months instead of 36 and the difference in return on an investment between construction costs of $22 a square foot and $35 a square foot.

His basic education had to come first as a member of the building team, in the tradition of learning to walk before taking up running. Some of this basic background came from attendance at meetings and seminars such as the National Conference of the Building Team. The Building Team conferences are a cooperative effort of major organizations in the construction industry:AIA, AGC, Building Owners and Managers Association, BRI, CEC, Mortgage Bankers Association, NECA, NSPE, and the Producers' Council.

The program of the Second National Conference of the Building Team, held in Houston May 10–12, 1972, stressed the following:

Time
Cost
Quality
Performance

These four words provide a simple summation of what it is all about. Today's designer/developer must respond to the owner's needs for shorter construction time, stable costs, quality in design, materials and function, and optimum performance in all respects. Sophisticated management teams must be formed now to fully utilize inherent abilities and resources.

It is imperative that all segments of the construction industry recognize this. Change is inevitable. None of us can afford to operate independently of each other. We must communicate. We must come together to discuss our common problems and opportunities. A team effort is needed.

As he heard more and more about 30-percent minimum returns on investment, with 50 percent not only desirable but often attainable, the design professional began to take a long, hard look at his own profits on fees of 5 to 7 percent. Some developers cleared more in one year than the architect or engineer might hope to make in ten years. Other developer-clients ended up pretty regularly in bankruptcy court. Some designers were put off by the attendant risks; others could not reconcile the constant emphasis on profits and ever-greater returns with their professional upbringing. But many came to watch and not a few have stayed to play—to take pieces of the action or even to generate the action. The remainder of this chapter will concern itself, for the most part, with the players and some of the rules (and pitfalls) of the game.

THE CONSORTIUM APPROACH

One approach to getting into the action is that taken by Ellerbe Architects in St. Paul. Possibly influenced by the results of a survey made for the Minnesota Society of Architects, indicating that 77.5 percent of building owners in the sample area turned first to a contractor instead of to an architect, Ellerbe initiated a design-build-own consortium in mid-1971. The comprehensive package offered by the consortium, known as Finance-Design-Build Associates, includes land acquisition and development, financing, construction, and operation of the project. Construction and permanent financing for the association are handled by a long-time Ellerbe client in the insurance field.

Ellerbe's consortium includes large independent mechanical and electrical contractors, but purposely omits a general contractor. This was done to enable the associates to capitalize on the special knowledge about such variables as politics, labor, codes, and community interests which a regional or locally based general contractor can bring to the team. Marketing efforts of the three permanent partners can be meshed for major projects—and in all cases, the individual marketing staffs should have an additive effect on each of the others' activities.

PORTMAN'S PROTOTYPES

John Portman, the Atlanta-based architect-developer, was certainly not the first design professional to take a piece of the action—but there can be little question that he is the most publicized of today's designer-entrepreneurs. Successful real estate development, according to Portman, is knowing the market, planning, timing, and merchandising. Further, the architect-developer must always keep sight of the parameters within which he can build something;

before construction begins, he must know what the potential is apt to be.

Following Portman's initial venture into real estate development —an esthetic triumph but a financial flop which reportedly cost him about $7,500—he took on the transformation of three run-down blocks in downtown Atlanta. Peachtree Center was the result; more than $125 million worth of office towers, a merchandise mart, a bus terminal and parking garage, and the Regency Hyatt House occupy the site so far.

The twenty-two-story merchandise mart was the first building in what was to become Peachtree Center. Battling and overcoming political and financial setbacks, Portman was able to interest such developers as Dallas-based Trammell Crow in his growing complex. Following completion of the merchandise mart, Crow and Portman went on to develop Peachtree Center as partners. David Rockefeller, Crow, and Portman are partners in the $200 million Embarcadero Center development in San Francisco.

Building on his experience in developing the merchandise mart, Portman decided downtown Atlanta needed a distinctive hotel. There was nothing particularly noteworthy or spectacular in that; designers are forever announcing that their cities need this or that kind of new building—and then waiting (usually forever) for a client to come along with enough money to build it. Here was where Portman parted company with his fellow professionals.

Forming his own development company, he designed and built what was to become the Atlanta Regency Hyatt House. If there was one thing John Portman did not want or need at that stage of his career, it was to own a distinctive, twenty-one-story hotel. He gambled that a buyer for the practically completed structure would be easier to find than a client for his plans to explore interior spaces within a commercial hotel. His gamble paid off when the Hyatt Corporation not only outbid other hotel chains for the Regency, but commissioned John Portman and Associates to do other hotels in Chicago and San Francisco.

SELLING THE CLIENT

The real dimensions of Portman's coup with the Atlanta hotel may be better illustrated with an imaginary presentation to a fictitious potential client:

ARCHITECT: Let's start with the lobby. We'll make it 110 feet wide, 140 feet long and, let's say, 220 feet high.

CLIENT: You've got to be kidding, Buster. That's enough cubage for 1,200 rooms, the way I build hotels.

ARCHITECT: But try to get the picture. There'll be nice, wide interior balconies running around each floor, overlooking the building-high atrium. You can have vines hanging down from all of the balconies — and how about suspending a bar from the roof and . . .

CLIENT: Hold it right there. For openers you threw away over a thousand guest rooms for an atrium — whatever that is. Now you want to make some kind of stupid greenhouse out of it with your nutty ivy plants — and what was that about a bar hanging from twenty-one stories up?

ARCHITECT: (pouting) Ed Stone gets to put atriums in his buildings. And he even hangs ivy out of the light fixtures sometimes.

CLIENT: You guys are all nuts. I've got a brother-in-law who knows this contractor, and . . .

Sensitive, experienced readers will have the general idea of how that particular conversation concluded.

The design-build-own bell has tolled in a variety of ways for other design professionals, depending on their size, individual desires, and salesmanship.

OTHER APPROACHES

Heery and Heery of Atlanta found a good marketing tool in guaranteeing both costs and on-schedule delivery of buildings to clients. If Heery and Heery cannot place a construction contract within their own cost estimates they lose the whole fee. When a contract is awarded, the contractor is penalized for late completion of the job.

When Charles Luckman Associates was acquired by the Ogden Corporation in January 1968, the shattering of professional design traditions was heard coast to coast. Ogden, a conglomerate selling everything from oceangoing ships to orange juice and hot dogs (Nedicks), has annual sales of more than $1 billion. Charles Luckman Associates was purchased for stock worth some $15 million at the time the deal was consummated.

Other firms, such as CRS Design Associates of Houston, have gone public to obtain the necessary financing to expand into new fields and establish branch offices. Smith, Hinchman & Grylls Associates, Inc. formed an entity with the power to initiate projects, secure tenants, and commission design. The corporation expects to have at least some ownership position in all of its developments. According to Robert Hastings, chairman of Smith, Hinchman & Gryll's board and a past president of the AIA, the firm elected to go into development for three primary reasons:

1. The economic advantages of ownership are attractive and give professional persons a means of building financial security.
2. There is long-term investment value to better-designed, better-constructed buildings.
3. There are some building types, such as housing and office facilities, where an increasing number of clients prefer the renting of finished products, and it is believed the development corporation will generate clients for such buildings.[5]

FARRELL'S RULES FOR DEVELOPERS

Paul B. Farrell, Jr., well-known consultant on developments, is a popular lecturer before architects, engineers, students, and others interested in breaking into the field of development. Farrell, trained as an architect, attorney, and urban planner, has formulated ten basic rules for profitable land development. First set down in an article for *Progressive Architecture* in March 1971, Farrell's rules for developers are:

1. Develop a "killer instinct."
2. Protect your position in a project.
3. Reject marginal projects early.
4. Don't confuse project cost with mortgage value.
5. Don't give up equity until absolutely necessary.
6. Know something about basic financing techniques.
7. Understand the real costs of borrowing.
8. Don't procrastinate.
9. Recognize lenders as unique personalities.
10. Surround yourself with the smartest people you can find.[6]

He later added rule 11: "In the real estate business—don't trust *anybody.*" Farrell's lectures include an explanation of how to analyze prospective development projects, illustrated by a simplified "Basic Real Estate Analysis Worksheet." The worksheet was carried in the *Progressive Architecture* article and has been widely distributed and republished since.

LEARN FROM OTHERS' MISTAKES

A more rhetorical approach to some of the points above appeared in an article in the *AIA Journal* for April 1972. Suggesting that designers on the verge of taking their first leap into development profit by the mistakes and financial setbacks of pioneering firms, the following general guidelines were given:

1. Start cautiously. Don't dive; wade into development work, constantly testing the water temperature. Development work entails financial risks along with opportunities. The architect who boldly plunges into the biggest game in town for his initial development project may find himself financially drowning. By limiting his equity participation to 20 percent of his professional compensation, an architect can assure that he will break even, or at worst suffer relatively slight losses on a development project. Yet this conservative investment, worth possibly 10 to 15 percent of the equity investment, greatly enhances his chances of controlling the design.

2. Do not compromise on professional compensation. Equity participation in a development project puts the architect's "blood on the line," in the words of Vincent G. Kling, FAIA. As members of development teams, architects, engineers and planners run the risk of having their contributions undervalued compared to those of contractors, developers, realtors, lawyers, etc. If the developer attempts to justify a low architectural fee as the price of an alluring equity-sharing opportunity, withdraw from the project. It may be the biggest game in town, but it cannot be the only game in town.

3. Ally yourself with competent, trustworthy partners. Set high standards for integrity. Consider even social compatibility with prospective co-joint venturers. A development project's co-owners are committed to a long-term association that is inevitably more intimate than the more formal architect-client relationship. It is one thing to have a so-and-so for a client; it is something else to be married to him as a business partner.

4. Attain at least some elementary comprehension of market and financial analysis and financing techniques. Learn, for example, how to make a review of the prospect's financial feasibility.

5. Investigate all aspects of equity participation. The architect should check into all these issues before getting involved; professional liability, possible ethical conflict arising from some peculiar aspect of the deal, financial investment prospects, etc.[7]

BASIC FINANCING TOOLS

The same issue of the *AIA Journal* just cited carried another article by the ubiquitous Paul Farrell, in which he covers briefly fourteen basic methods of real estate financing, representing "over 99 percent of all real estate financing transactions."

Readers are referred to the Farrell article for his definitions of the following basic tools of real estate financing:

1. Construction (interim) financing
2. Permanent first mortgage loans
3. Gap financing
4. Standby commitments
5. Leasehold financing
6. Land sale-leasebacks
7. Sale leaseback of energy system
8. Sale leaseback with 100 percent financing
9. Installment sales contract
10. Bonded lease financing
11. Wraparound mortgages
12. Secondary financing
13. Purchase money mortgages
14. Development loans[8]

If nothing else, this list should clinch Farrell's earlier advice to those contemplating getting into development: "Recognize lenders as unique personalities" (Rule 9), and "Surround yourself with the smartest people you can find" (Rule 10).

IS IT FEASIBLE?

The usual two-phase test of the economic feasibility of any development project consists of the market analysis (a la Farrell's "Basic Real Estate Analysis" or a similar approach), plus the construction analysis; or, how much will the project cost to design and build?

The primary question for a developer is whether or not he can provide the equity money over and above the amount which can be financed (front money). Front money is the difference in dollars between the amount which can be borrowed by the developer and the funds required to bring the investment up to 100 percent of the total project cost. A shortage of front money is responsible for the killing of about 95 percent of all construction loans that fall through.

After a careful calculation, if it is apparent the front money cannot be raised, forget the project. There are a few lenders who specialize in loaning venture capital, but at very high rates of interest.

WHERE TO BORROW

Certain professional men, such as physicians, attorneys, dentists, accountants—even a smattering of architects and engineers—can be a source of front money. They often are interested in finding depreciation benefits for their personal income positions. Occasionally, professionals will loan venture capital at lower than the prevailing commercial rates. Some development projects have an obvious appeal. For example, investors who go into a conventionally financed apartment in the preconstruction stage can share in tax write-offs from both development and operations.

Another source of front money is in the form of a line of credit from a commercial bank. This is often based on the developer's successful operation of another—even unrelated—type of business. Ask for an open note in the form of an irrevocable and unconditional line of credit—not a loan. The bank charges a fee for this, of course, but it is well worth the fee in that a person's credit, rather than his money, is working for him.

One alternative, when faced with a shortage of front money, is to syndicate the project; i.e., take in partners who will furnish cash or services in exchange for an equity in the project. For example, the landowner might come in for a share based on the appraised value of the land, or the architect could leave his design fee in the project.

HOW TO BORROW

After he is assured of sufficient front money the next thing a developer must do is to arrive at a conservative estimate of the cost of his project. Lenders look at total costs, including nonconstruction costs such as interest, finance charges, land costs, start-up, and operational costs (as in a motel). These costs are very significant to a construction lender in arriving at a decision as to how much to loan, if anything.

In the halcyon days of the early 1960s almost anyone could get a 100 percent loan of the true costs of a project—mortgage out, in other words. Today, 90 percent of true costs is about the best one can do—and this ordinarily is only if the developer owns the land.

Assuming the developer has his costs firmly in mind, there are a few other details he should be prepared with before approaching a construction lender. As a minimum, the developer should have a reasonable floor plan and design from the architect to show the prospective lender. Ordinarily, he can get a loan commitment with just that much, although he will get no money from the construction lender until he has a full set of plans.

The developer should have some type of feasibility study, for his own sake as well as to show the lender. Utilities and zoning should be checked, the political climate evaluated, etc. Unexpected zoning changes and sewer moratoriums could have a disastrous effect on the project. From a study of comparables the developer should know what rents he can reasonably expect. Some construction lenders require an independent appraisal before committing a loan, setting forth what the project would bring if completed and sold on the open market. A good appraisal will include a feasibility study.

In addition to the plans and feasibility study, the construction lender will take a close look at the borrower (developer), his construction track record, character, and financial strength. The developer should also bring a current (not more than six months old) financial statement and a list of credit references to the interview. Include all possible bank references. If the developer is a contractor he should have a list of subs and material suppliers with whom he has dealt. Furnish a fairly detailed résumé of your background and experience, completed projects, type, location, etc. The lender will look at them before he commits. It is essential, unless one is looking for a package deal, to bring along the permanent loan commitment. The permanent loan is always obtained first—otherwise a construction lender will have no interest in talking to you.

If he does not have an established track record in development, then the potential developer will have to sell himself. On a first loan, with no record, he will be lucky to get a loan of more than $750,000, although the amount of his equity enters into the decision of how much to loan. The inexperienced developer can (and should) surround himself with experienced advisors and consultants. The lender will look at the caliber of the advisors.

A reference recommended by several experts in the development field is *The Real Estate Desk Book* by William J. Casey. Now in its fourth edition, the *Desk Book* is published by the Institute for Business Planning, New York, N.Y. The prolific Mr. Casey has written desk books on a number of business and financial subjects, so make sure you get the one on real estate.

WHERE IS THE ACTION?

There is no practical limit on where or how a piece of the action may be found or generated by an alert, imaginative, and development-minded professional. As an example, consider the leisure-time industry through the foreseeable future. Some economists predict that the leisure market will reach $250 billion by 1975, up from $160 billion in 1970. Over the past 100 years the average American worker has gained something like 50,000 additional leisure hours during his lifetime. These almost six extra years of freedom from work per working man and woman have already made the leisure market a profitable one for wise investors and developers, in everything from mobile homes to camping gear to sports equipment.

ACTION IN FRANCHISES

The franchising industry, which is somewhat related to the leisure market, also saw a remarkable growth in the sixties and early seventies. Of all the franchised businesses available to a man with several thousand dollars, a large family, and a will to succeed, perhaps the most popular have been the fast food service shops. Some of these have fallen on hard times as franchisers overextended and overbuilt, but there always seems to be room for a new idea and another approach.

There is no reason why a design professional cannot come up with an idea which could be licensed or franchised—and some have. After several years of cultivating indigestion on the run from such roadside culinary delights as fried chicken by the basket and hamburgers by the sack, perhaps American eaters are ready for a little atmosphere, imagination, and class in their eating habits and surroundings. There are, of course, other kinds of adventures in dining out besides upset milk shakes in your lap, French fries in your pockets, and ketchup in your hair.

For the purposes of developing a fictitious case history of how a franchise idea evolves, let us consider fondue, a French word meaning "melted." Fondue dishes, based on those served in

France and Switzerland, have gained wide popularity in the United States—but mostly as at-home preparations. Meat or cheese fondues serve as main courses, with chocolate fondues substituting for the more traditional desserts.

"FONDUE HAUS" FRANCHISE

A designer-entrepreneur on the East Coast decided that the already established acceptance of these one-dish meals, coupled with a promotion campaign, might be turned into a profitable franchise operation. After a quick swing through Switzerland and Bavaria to pick up ideas, it was no problem for him to design representative interior and exterior treatments for a "Fondue Haus" franchise. Naturally, he worked closely from the outset with an attorney experienced in franchise matters. The attorney handled such items as incorporation and contracts and registration of trademarks, trade names, and business marks in the several states where the operation was to begin.

The plan was to offer restaurant franchises in selected high-traffic locations, featuring fondue dishes. The menu was built around six basic fondue dishes, two cheese, three meat, and a chocolate dessert fondue. The exterior plan had to be flexible in order to meet local requirements, established design concepts in shopping centers, frontage dimensions, and other considerations. The interior design was based on a small *Gasthaus* (restaurant) in the German-speaking section of Switzerland, but brought up to modern standards. Heating elements for the fondue pots, for example, were incorporated into and sunk slightly below the table tops—for safety and so that insulated covers could be put over the electric elements when they were not in use. Where licenses were available domestic and imported wines and beer were also offered. A profitable sideline was the sale of fondue recipe books, pots, and prepared fondue mixes and sauces. One of the benefits of this type of food service is that there are relatively few dishes to wash.

At the proper time the developer's attorney brought a franchise consultant into the picture. Within two years more than twenty-five Fondue Hauses were in operation from Boston to Washington, D.C., and the design professional who originated the idea was living very well from the gallons of fondue ingested daily by happy customers. The bad news in all of this was that he could no longer stand even the sight of a bubbling fondue pot—but few fortunes are made without some undesirable and unexpected side effects.

DISREGARD THE OBVIOUS

Architect O'Neil Ford advises his fellow professionals to "develop a thorough disregard of the obvious." Ford illustrates his counsel with the story of a new supersonic plane under development in a large aircraft factory. The prototypes kept failing their test flights because of a structural failure in the starboard wing. No matter what modification the engineers came up with, the right wing always failed in the same place. During the critique following the latest crash a long-time employee, now semipensioned off in a janitorial position, spoke up to suggest that a line of holes be drilled along the fracture line on the next test model. He remembered that holes redrilled along a line of weakness interrupted incipient fractures. In his current work as a janitor he was reminded of this fact because he noticed that toilet paper seldom tore along the lines of perforation.

Edward Durell Stone's famous and often-published rules for designers to live by include suggestions such as the following, in rule 5: "Don't fall in love with your first idea. If you are jealous and overly-possessive of your ideas it is a sign that you have too few. Keep an open mind—the janitor's suggestion is probably the best."

REFERENCES

[1] *From a Dictionary of American Slang,* by Harold Wentworth and Stuart B. Flexner, Copyright © 1967, 1960 by Thomas Y. Crowell Company, Inc. P. 388. With permission of the publisher.

[2] Gurney Breckenfeld, "The Architects Want a Voice in Redesigning America," *Fortune Magazine,* November 1971, pp. 144–145.

[3] Ibid., p. 145.

[4] R. Buckminster Fuller, "The Age of Astro-Architecture." Copyright © 1968 by Saturday Review, Inc. First appeared in *Saturday Review,* July 13, 1968, pp. 17–20. Used with permission.

[5] The *AIA Journal,* Washington, D.C., May 1971, p. 39.

[6] Paul B. Farrell, Jr., "Ten Rules for Profits in Land Development," *Progressive Architecture,* Stamford, Conn., March 1971, pp. 88–90.

[7] C. W. Griffin, Jr., "The ABC and Why of Development Building," *AIA Journal,* April 1972, p. 37.

[8] Paul B. Farrell, Jr., "Basic Real Estate Financing," *AIA Journal,* April 1972, pp. 38–40.

CHAPTER 14

Public Relations: Basic

For the purposes of this book the subject of public relations has been divided into two chapters. The present chapter deals with some of the basics: history, definitions, general techniques, the current state of the art, and similar matters. Chapter 15 covers public relations for the design professional, with case histories, specific applications, and (hopefully) a clear and cogent explanation of how a productive public relations program interfaces with and strengthens an effective business development plan.

The field or profession of public relations is an everexpanding one, with increasingly important implications for the marketing of professional design services. To state the obvious: Any subject which has more than 950 books about it listed in the card catalogue of the Library of Congress cannot possibly be covered completely in one or two chapters, or even one or two volumes. The treatment here, of necessity, must consist of little more than a survey course.

The 950-plus listings in the Library of Congress do not include books on such closely related subjects as publicity, advertising, effective writing, and public affairs, so the grand total of the printed word on the general subject conceivably runs to several thousand volumes in at least seven languages. The range of books available to the student of public relations is indicated by the first and last catalogue entries. *Public Relations; Theorie and Systematik,* by Carl Hundhausen (Berlin, 1969), leads off; *Whither Public Relations Work,* by William A. Hamor (Pittsburgh, 1935), is the final entry. One wonders if the cataloguers were making a sly social comment when they set up the subject heading following "Public Relations" as "Public Relief."

With such a wealth of available material on the subject, the layman's confusion as to where to begin his self-education is understandable; some confusion on the part of the PR professional is also excusable. The Public Relations Society of America, the professional association for those in public relations, has a program of professional accreditation for its members. PRSA Accreditation bears some similarities to state registration for the design professional but has no legal standing, and no one is prohibited from offering himself to any client as a public relations counsel without accreditation. To help members prepare for the accreditation examination, PRSA recommends a number of reference sources. The following list of suggested books is taken from that list:

Effective Public Relations, 4th ed., Scott M. Cutlip and Allen H. Center; Prentice-Hall, 1971.

Handbook of Public Relations, 2d ed., Howard Stephenson, ed.; McGraw-Hill, 1971.

Crystallizing Public Opinion, rev. ed., Edward L. Bernays; Liveright, 1961.

The Management of Promotion, Edward L. Brink and William T. Kelley; Prentice-Hall, 1963.

Public Relations: Principles, Cases and Problems, 5th ed., Bertrand R. Canfield; R. D. Irwin, 1968.

Lesly's Public Relations Handbook, Philip Lesly, ed.; Prentice-Hall, 1971.

Public Relations Handbook, Richard W. Darrow, Dan J. Forrestal and Aubrey Cookman; Dartnell, 1967.

Only By Public Consent, L. L. L. Golden; Hawthorn Books, 1968.

The Nature of Public Relations, John E. Marston; McGraw-Hill, 1963.

HISTORY

Public relations has come a long way, in its own way, since the turn of the century. Ivy Lee is usually credited with inventing public relations, although there is another school which holds that Edward Bernays was the father of modern PR. The latter school included Mr. Bernays. Nevertheless, both Lee and Bernays had much to do with shifting the concept of public relations away from press agentry, along with elevating the position of a public relations man to counselor status and the first vestiges of professionalism.

In 1910 the Manhattan telephone directory listed not a single public relations counselor. Fewer than thirty PR counselors appeared as late as 1935; in 1972 almost 900 practitioners were listed in the Manhattan classified section. Although reliable statistics are difficult to come by, educated guessers estimate that between 50,000 and 100,000 persons are actively engaged in public relations in the United States today. The Public Relations Society of America has a membership of something more than 6,000 individual and corporate practitioners, or between 6 and 12 percent of the potential.

A *Time* magazine "Essay" in 1967 had this to say about modern public relations:

> . . . the good public relations man is more than a press agent—though not even the best is ever wholly free of flackery—and considerably less than Big Brother. His calling contains more than its share of what the *Nation* long ago called "higher hokum." But it is also a legitimate and essential trade, necessitated by the complexity of modern life and the workings of an open society.

The primary concern of the early PR man or "press agent" was press relations, but the modern PR counsel has a greatly broadened field of responsibilities and activities. These include the traditional press relations function, but also encompass investor and shareholder relations, public affairs, educational and minority relations, civic affairs, politics, and several other equally important areas. Today's effective PR man must be a specialist in communications. He is no longer strictly a pitchman, although there are many elements of salesmanship involved in the normal and accepted practice of public relations.

Robert Newman, director of corporate relations and public affairs for the diversified conglomerate, TRW, Inc., uses this definition of an effective PR counsel:

Good PR men must know what to speak, when to speak, where to speak and how to speak—when to listen and when to keep quiet.   They are not—nor should they be—a buffer between the company and the various publics interested in the company and of interest to the firm.   They are—and should be—an interface, often a catalyst, between management and the shareholder, the potential investor, the student or engineer looking for a job, the customer in search of a product (or service), the general public and the various levels of government under which the company is a corporate citizen.   The chief function of good corporate PR men today is to convey understanding of the company, its operations, its objectives, its philosophies, its good points—and bad.

While Newman was describing a corporate PR man for a firm with more than 80,000 employees and consolidated sales of over $1 billion in 1967, many of his points relate to every PR practitioner.

DEFINITIONS

Formal and informal definitions of public relations run the gamut from idealistic to gutsy, from the sublime to the ridiculous.   An example of the latter is the PR man who became overawed by his own importance and insisted that his craft was "social engineering."   Another definition of public relations which enjoyed transitory favor was "the engineering of consent."   Not only did it sound a little dirty but it had other undesirable connotations and actually implied more than public relations is able to deliver.

A few other definitions from various sources:

- Let your light so shine before men that they can see your good works.   Matthew 5:16.
- Public relations: Performance plus communications equals reputation.
- Public relations is the nasty letter you don't write when you're mad—and the nice letter you write the SOB the next day after you've regained your sense of humor.   Conversely, PR is the nasty letter you write when you're mad, then tear up because you've gotten 95 percent of the hostility out of your system by just writing it.
- Public relations is the management function which evaluates public attitudes, identifies the policies and procedures of an individual or organization with the public interest, and executes a program of action to earn public understanding and acceptance.   *(Public Relations News)*
- A formulation: Public Relations = PR = Performance and Recognition.
- Another formulation: T^3PR (Try to Treat People Right).

- Good public relations is the art of not treating the public like relations.

- Public relations is the business of the Invisible Sell, or Dale Carnegie Writ Large; the professional winning of friends and influencing people—not for one's self but for one's client.

- The deliberate, planned and sustained effort to establish and maintain mutual understanding between an organization and its public. (British Institute of Public Relations)

On the other hand, there are many things public relations is not: "My public relations policy shall consist of a maximum of politeness and a minimum of information." (Reportedly stated by Mrs. Jacqueline Kennedy Onassis to her White House staff.)

According to one PR man:

> . . . professional public relations is NOT press agentry for some buxom new movie starlet; neither is it fake publicity for a product of questionable value, nor propaganda for questionable causes, nor advertising, nor sales promotion, nor personal appearances at the ladies' garden club, nor lobbying in legislative corridors, nor boozing and feeding news media reporters. All of these activities—publicity, propaganda, advertising, marketing, speeches, lobbying, media relations—may at one time or another be part of a specific public relations program. But taken singly, or in some context other than the PR objective of providing a more favorable environment through attention to the enhancement of favorable public attention, so that a business, organization or institution can operate more effectively, these activities are NOT professional public relations—and the person whose job is exclusively in any one of those areas is NOT a professional public relations practitioner.

The above negative definition of public relations (what it is not), raises a long-standing point of contention among PR professionals, the individual and collective perimeters and relationships of such matters as public relations, publicity, propaganda, advertising, and sales promotion. Since hundreds of master's theses and doctoral dissertations have covered the finer points of each of these elements, our treatment of them will not attempt to go into extensive detail.

Public relations, publicity, advertising, and sales promotion are all basic parts of the successful marketing techniques of American business. Propaganda, as a highly specialized form of publicity and not overly germane to this discussion, we will ignore.

Marketing deals with the sale of products and services. Products and services, because of their different natures, must be marketed in different ways. It is obvious that a professional service cannot be sold in the same manner as a breakfast food or a lipstick. This is where the concept of public relations and the techniques of publicity come into play.

Public relations for the architect, engineer, or planner means getting to be known, liked, trusted, respected, and preferred — while developing and upholding professional dignity. Public relations is not witchcraft; it cannot make something out of nothing. It merely points out the deserving person to the public. The public cannot be blinded to nefarious deeds by anyone or any organization, nor can fine words substitute for honest acts. The late Carl Byoir, one of the PR profession's earliest and most successful practitioners, once said: "It is absolutely impossible to whitewash a manure pile. Nothing will help but to shovel the stuff away and start from scratch."

There is an old PR fable which relates to Byoir's inelegant but eloquent observation. A tiny bird was both freezing and starving to death on a lonely road in Siberia. A peasant, seeing the bird's plight, picked it up, nestled it in a manure pile beside the road, and proceeded on his journey. The warmth soon revived the little bird. The unusual "nest" provided nourishment, too. Before long the bird was chirping and singing merrily. A second peasant came along and marveled at the bird's high spirits in such a plight. Quickly and gently he removed the bird from the manure, carefully brushed it off, and set it back down on the road. The peasant went on his way. The bird froze to death.

There are three important lessons to be drawn from this fable:

1. It isn't always your enemies who get you into it.

2. The person that gets you out of it isn't necessarily your benefactor or friend.

3. When you're in it up to your neck, have the good sense not to sing about it.

The public relations counsel, it might be pointed out, is not retained as a substitute for company thinking and action on promotion, but rather for his experience in deciding what is or could be made news and to stimulate and direct the firm's thinking and initiative in the field of promotion. The counsel's function may be likened to that of a combination catalyst-sifter-funnel-amplifier, plus the sometime role of devil's advocate. Professional, productive public relations is always a cooperative effort, a two-way street between client and counsel.

Anything that is news to people is publicity, since publicity basically is the reporting of newsworthy events. News has been defined as anything which excites, astounds, amuses, or informs the reader, listener, or viewer. Here is where the professional public relations counsel begins to earn his fee, since it is his business to know what really is news, or, more important, how to develop a news angle or "peg" to help sell an otherwise ordinary or even uninteresting story. Never make the mistake of confusing publicity with public relations. You have public relations whether you want it or not, but one is usually free to choose whether or not he has publicity.

Advertising, conversely, is paid space pure and simple. Space, in this context, includes radio and television air time, as well as newsprint. Ordinarily, anything within the bounds of good taste and legality may be stated or pictured in that space, whether it is news or not, and remains under the full control of the advertiser.

Sales promotion is dedicated to improving sales and business. It may be thought of as the blanket that covers all efforts to increase one's business, and includes advertising, market research, publicity, direct mail, and personal solicitation. In at least some of its forms sales promotion can be applied either to a product, such as an automobile battery, or to professional services such as architecture and interior design. The strength and success of the total sales promotion effort depends upon the effectiveness of whichever of its components are employed at any given time.

Public relations is also the practice of *evaluating* a firm's services in relation to public needs and wants, *identifying* services with those wants and needs, and then *communicating* the resulting ideas to the public. Any public relations program of any size and purpose can be measured against these three conditions.

Evaluation. This requires an analysis of what services the firm is prepared to render and what new services it should prepare itself to offer in view of public demands. This, in turn, requires an examination of the client's completed projects; his opinions on community, state, and national—even international—problems; his memberships in community and professional organizations; his office and staff. The sum total of all these factors will normally be a reflection of the firm's professional and practice principles.

Identification. Properly handled, the analysis can be converted into speeches and articles, releases and photographs, brochures and presentations, all of which will serve to relate the firm's work to the public need.   Identification is the key to success in any public relations program.   You can only interest people in what you do, show, or say by putting your message in their terms and into the specific areas of their interests.   The transmitter and receiver must be on the same frequency for the message to get through.

Communication.   Having established a common frequency, it remains for the firm's public relations counsel to choose the appropriate communications media—publicity, civic activity, radio and television appearances, office activities, etc.—for each particular occasion.   A sense of timing, knowledge of relevant issues, and familiarity with the personalities involved are all important considerations in choosing the proper medium or combination of media.   This sense is developed through experience, which in the final analysis is what the design professional is buying from a public relations counsel.

As was pointed out above, public relations and marketing are integral and essentially inseparable elements of a total sales program.   Where a corporate organization does not put both public relations and marketing under the same executive overview, channels for close coordination and cooperation are always established. Some larger design offices follow the same pattern.   Ellerbe Architects, for example, has a Director of Marketing and Communications.

PR MASQUERADERS

The experienced public relations professional is painfully aware of the growing number of individuals and enterprises who masquerade as PR counsels.   Some members of this fringe group are legitimate; many are not.   Former political office holders and retired athletes are especially prone to use the term "public relations" to describe their new careers.   Few of the PR masqueraders bother with training or experience in their new craft.   Possibly the best (or worst) example of the misuse of the term is provided by the postconviction statement of a Kansas City, Missouri, newsstand operator, who was found guilty of peddling obscene movies from his newsstand cover on a busy downtown corner.   Since it was his first conviction, a lenient judge released the news dealer on parole. Asked by a newspaper reporter what his plans for the future were, the convicted seller of pornography said he would probably "go into public relations" because he "liked people."   Whatever their shortcomings may be, the state registration laws do preclude the entry of untrained and unqualified persons into the design professions because they "like buildings."

A reason for the rapid growth of public relations — both qualified and unqualified — is the vastness and complexity of our modern communications complex. In the United States alone we have four major national TV-radio networks, more than 6,000 radio stations, and some 800 television stations. Add to those opinion-forming media more than 20,000 newspapers, magazines, and trade publications; sponsored films; and direct mail pieces and newsletters on practically every subject known to man, and you have some idea of the information cacophony unleashed daily on America.

Alvin Toffler, in *Future Shock*, observed that ". . . there are discoverable limits to the amount of change that the human organism can absorb, and that by endlessly accelerating change without first determining these limits, we may submit masses of men to demands they simply cannot tolerate. We run the high risk of throwing them into that peculiar state that I have called future shock."

In noting the rise of irrational responses on the political left as well as on the right, author Toffler suggested that one predictable maladaptation to an overload of inputs (future shock) will be for the beset individual to blow a fuse — to refuse to take in new information and to build up mental barriers which, when finally breached, will probably lead to personal catastrophe.

An easily understood example of what Mr. Toffler describes is the difference in information available to the peasant living in the Middle Ages and that pressed on today's average American. Historians and communications experts point out that modern man is subjected to more information from all sources in one day than the medieval man was exposed to in his entire lifetime.

Philip Lesly, Chicago PR counsel and author, summed up the problem in these words:

> Things were simple when forming a committee or issuing a statement created public opinion. They were still simple when press conferences or junkets were the focal point of communication efforts. Today and tomorrow the seething currents of human aspiration, misconception, alienation — all the passions of a people who are freer of life's perils than ever and therefore are restive with the few disciplines that remain — are the underlying basis of what we must cope with. For communications media and for professional communicators, that is the challenge of our segmented society.[1]

Not long ago a respected New York PR counsel, Farley Manning, compiled what he called "A Dozen Thoughts on Communicating With People." The paper soon became known as "Manning's 12 Communications Commandments" within the public relations profession. By whatever title, the list is well worth reading—and remembering.

1. The truth provides a ceiling above which public opinion cannot be maintained.

2. If you present a person with more than one idea at a time, he will probably retain only one—and it may not be the one you want him to retain.

3. The easiest way to reach people is in the context of their own special interests.

4. People tend to rate others on the basis of their own special interests.

5. People tend to believe "inside information" over public statements.

6. People don't really read or listen. They see or hear only what they want to see or hear. (Or reject what they do not agree with.)

7. Third party endorsement is extremely valuable from a respected source—but it can work both ways.

8. For a subject to hold the attention of the public, controversy must be involved.

9. There are many layers of opinion leaders and almost never can you reach all of the people in one step.

10. Logic practically never influences anyone—but it is a big help in justifying emotional appeal.

11. People rarely remember facts, but repetition of facts (or unrefuted fallacies) tend to establish attitudes.

12. Never leave a false statement unchallenged or it may become a part of the literature.

PUBLICITY

It was pointed out a few pages back that organizations and individuals have public relations whether they want it or not, but publicity is usually gotten by choice.

Publicity has been defined as "telling the story." As we have already seen, publicity is news and sometimes vice versa. Publicity traditionally comes in two forms, manufactured and natural, or planned and accidental. Planned publicity includes such special events as groundbreakings, topping out a building, and dedications.

Herbert M. Baus, in *Public Relations at Work*, gives a list of characteristics for determining the "newsworthiness" of an item, with examples of each:

1. When it is news; the atomic bomb.
2. When it is novel; Hirohito waits on MacArthur.
3. When it relates to famous persons; any Hollywood column.
4. When it is directly important to great numbers of people; information about the income tax.
5. When it involves conflict; battles, divorces, athletic contests.
6. When it involves mystery; most crimes.
7. When it is considered confidential; the revelations of some columnists.
8. When it pertains to the future; plans for improving a city.
9. When it is funny; Jim Moran personally hatches an ostrich egg.
10. When it is romantic or sexy; weddings or any feature story and picture of a pretty girl.[2]

One must remember that Baus's examples are more than twenty-five years old—thus accounting for mention of the atomic bomb as "new" and the reference to Emperor Hirohito and General MacArthur. The Jim Moran mentioned in characteristic number 9 was a free-spirit press agent, known for creating such manufactured publicity-news items as selling an icebox to an Eskimo and becoming the father (mother?) of a baby ostrich by setting on the egg until it hatched.

Remember the definition of news: anything which excites, astounds, amuses, or informs the reader, listener, or viewer. News is also apt to be based on deeds rather than words; specifics rather than generalities; benefits rather than features; and it is practically always related to known public interests (recall Manning's Communications Commandment number 3).

Four rules for handling publicity were given in an article in the *Public Relations Journal*, "Is Publicity a Neglected Art?"

1. Find the story. No reporter ever did a good job just "taking news" he got over the phone and the public relations man has to dig even harder.
2. Determine the story's worth to the company, editor and reader. This is not as simple as it sounds. There are compromises all the way along the line.
3. Think the story out, exploring every angle, twist, interest level, etc. Avoid the mechanical "bored reporter" approach that gets the facts right, but always looks like filler copy.
4. Shape the release, selecting tone, language and structure according to the subject matter and various editorial interests. There are so many variables determining the way a release is written that any "rule" is dangerous, but, in general, an exciting style helps any release as long as the "excitement" is balanced by "content."[3]

The same article concludes:

> If there's one place in this exasperatingly imprecise field of endeavor [public relations] where a professional really faces a constant challenge of performance, it's in publicity. If there's one place where there's no substitute for sheer skill, it's in publicity.
>
> When you are working on a story for a new group there is an inevitable skepticism if the people involved aren't familiar with your work. They don't really think you can do it. It's just a bit like being a pool shark in a strange town. They don't know it, but you do. You're going to clear the table and you can do it every time.[4]

In view of the importance placed on publicity and the writing of news releases which editors will use, the results of a mid-1972 survey of public relations graduates must have been a little chilling to all public relations professionals and not a few of their clients. Asked to list the areas they felt least prepared for academically, the newcomers to PR indicated they lacked training in photography and graphics, general business practices, and writing-editing, in that order.

SPEECH WRITING

> "Except ye utter by the tongue words easy to be understood, how shall it be known what is spoken? For ye shall speak into the air."
>
> I Corin. 14:9

A young speech writer once pointed out the difference between the epistles and the gospels and their relation to effective speech writing.

> The epistles are letters from the apostles to their brethren warning them against sin and urging them to do good; little sermonettes which most of us need—but abstractions.
>
> But the gospels are stories. Nothing abstract about them. They start out like stories, and anytime you come across an opening sentence like: "Now in the fifteenth year of the reign of Tiberius Caesar . . ," you know that something is going to happen.
>
> Most people can't remember much from the epistles. Everyone remembers the gospels. That taught me a lesson. If I had my way, every speech would be a succession of stories. As it is, I go to a great deal of trouble hunting actual examples to illustrate the points I make.[5]

One of the reasons for making a speech is the possibility it will have publicity value beyond the audience present to hear the address. If the speaker or his audience is important enough, just the fact that a speech will be given is often worth an advance story.

Since the number of speeches being given is rising sharply every year in the United States, it becomes more difficult by the day to get publicity mileage out of a platform appearance. A few years ago it was estimated that 20,000 speeches a day will be given in New York City by 1978. An average fifteen-minute speech uses around 2,000 words, so New Yorkers may be sitting through some 40 million words daily in a few years. Small comfort may be taken from the fact that some human minds can take in as many as 600 spoken words a minute, but even a fairly rapid speaker seldom delivers more than 200 words a minute.

A few other statistics on words and spoken communications: An unabridged dictionary has more than 600,000 words, plus some 150,000 more technical terms. For the 500 most-used English words, the *Oxford Dictionary* lists 14,070 separate meanings — an average of twenty-eight meanings per word. One hundred of the world's most popular proverbs use only 650 words out of the more than 750,000 available: less than one-tenth of 1 percent of the potential.

GHOSTWRITERS

The main differences between a good and a bad speech are:

1. Research
2. Preparation
3. Delivery

A good ghostwriter can take care of the first two items and save a busy speaker the days of research and writing required for an effective presentation. Working productively with a ghostwriter requires several things of the speaker who will use the ghost's efforts:

1. Particularly on the first two or three speeches, take time to talk at length with the writer. He needs to get a feel for the vocabulary, ideas, and platform habits of the speaker. By the time a dozen or so speeches have been successfully ghosted and delivered, this should be second nature to both parties.
2. Have the writer give you a list of potential questions which might be engendered by the talk — along with the answers.
3. As further protection on the dais, ask the writer to document all statements, in case you get a question from the audience about the source or subject matter.

If you cannot afford a good ghostwriter or do not believe you could be comfortable with that approach to creative speech writing, there are other avenues of assistance open to the busy design professional. For only ten dollars a year, texts of speeches by the world's best-known speakers—many of whom employ their own expensive ghostwriters—will arrive twice monthly in magazine format. A file of back issues of *Vital Speeches of the Day*, published by the City News Publishing Company, Inc., Box 606, Southold, New York 11971, could be worth a dozen ghostwriters. An annual index to the speeches is carried in a November issue; for $15 the publisher will send a cumulative index covering volumes 1–25.

WRITING

All of us make contacts, just as a salesman gets to know his customers and would-be customers, and sometimes, just sometimes, a friend will at least listen to your story.

But contacts alone won't get you into print. Or on the air. Or on a TV show. Space and time are too dear. An editor who puts together his show or publication largely on the basis of the pleas of friends wouldn't last long in the highly competitive numbers world of audited circulation and ratings. His output would soon sicken. The medium with which he had heretofore charmed his audience would soon wither and die.

And so, as with any other organization, we must *sell* our wares. And THE PRODUCT MUST BE DESIRABLE.

No question about it. In Communications THE STORY (our PRODUCT) IS THE THING.

Story excellence, of course, implies that all of our writing must be of high quality. The subject matter must be made as exciting as it can be, as well as CLEAR. And clarity is an absolute must! If it can't be made clear, don't tell it!

After clarity, then, make your story BRIEF and make it sing. It will be in competition with the daily, weekly and monthly output of tens of thousands of competent writers and editors on countless staffs of innumerable news- and feature-producing organizations—plus the output of free-lance contributors the world over . . . and, oh yes, those tens of thousands of publicity men all fighting to draw the editor's attention to their wares. All of these people are striving to lay claim to that precious, but limited, space and time that make up our communications media.

Granted, you're not competing with *all* of them at any given time or with any given offering. But, rest assured, there's a stack of stories at every editor's elbow, all begging for use.

One thing that will help your story immeasurably as it intrudes on an editor's busy day is its appearance. It should fairly shout to the editor into whose hands it falls: "I'm the kind of product you're used to dealing with. I'm familiar to you. Pick me up and read me. You'll find something here that your readers want to know."

So STYLE, too, IS THE THING.[6]

The above is from the foreword of one of the best general guides (stylebooks) on writing releases, published for the staff of the public relations department of Xerox Corporation. Copies are not generally available to outsiders, but if an opportunity to beg, borrow, steal, or duplicate a copy of it ever presents itself, you will find it of great help in understanding the sometimes mysterious ways of wordsmiths, publicity writers, and the entire public relations genre.

Other excellent stylebooks are more readily obtained by the layman. These include the stylebooks used by UPI, AP, and the *New York Times (Style Book for Writers and Editors)*. The latter is published by the McGraw-Hill Book Company. A much more detailed approach is the University of Chicago Press, *A Manual of Style*.

Publicity writing usually has a point of view. This is not to say it is necessarily slanted in favor of the client but the warts are usually covered up or ignored. Illustrative of "point-of-view" reporting are the following translations of successive headlines from a Paris paper, which appeared over stories about Napoleon's escape from his first exile on Elba:

THE CORSICAN MONSTER HAS LANDED IN THE GULF OF JUAN

THE CANNIBAL MARCHES TOWARD GRASSE

THE USURPER ENTERS GRENOBLE

BONAPARTE ENTERS LYONS

NAPOLEON MARCHING TOWARD FONTAINEBLEAU

HIS IMPERIAL MAJESTY EXPECTED TOMORROW IN PARIS

Herbert Baus, in his work cited earlier, illustrates the power of one word to color fact. Baus uses the example of a state governor refusing to see a delegation of communists. For our purposes, we will make it a delegation of student activists from the state university. The incident, Baus points out, might be reported in any of the following terms—in each case, the use of a single adjective giving a distinct coloration to the plain fact: "The governor refused . . ."

<table>
<tr><td></td><td>frankly</td><td></td></tr>
</table>

	frankly	
	stubbornly	
	bluntly	
	weakly	
	bravely	
	sourly	
The governor	blithely	refused
	impatiently	
	hastily	
	shrewdly	
	cautiously	
	timidly	
	foolishly	

PHOTOGRAPHS

Almost every story about people and most stories about things are enhanced by a good photograph or two. According to Mel Snyder, PR man and free-lance photographer, there are three very good, simple reasons to use photographs in a publicity program: "to communicate information quickly; to create excitement without easily recognized bias; and to offer graphic proof of intangibles such as market acceptance, style leadership, quality production, modern manufacturing techniques and dynamic management."[7]

Advance planning, an occasional change of angle, and intelligent cropping are all important in getting the right kind of publicity shot. Good storytelling photographs must be technically perfect and make a definite contribution to the story.

A smaller portion of the full negative often will tell a more dramatic story than the complete film image. The widely published picture of President Kennedy's son John saluting as his father's funeral cortege passed by was less than one-fifth of the original photograph. Competent picture editors are not afraid of a little photographic surgery to achieve visual impact and tell a better story.

TV IN PR

A 1971 Roper poll found that 60 percent of respondents relied on television as their first source of information. In answer to another question as to which is the most believable medium, 49 percent said television against 20 percent for newspapers. (Perhaps the most significant statistic from this survey is the almost one-third of the respondents who apparently do not believe either one very much.)

Nevertheless, no public relations counsel can afford to ignore the selling power of the tube. A fairly hot item in the last few years has been the television news release. The *Wall Street Journal*, no particular friend of public relations and its practitioners, called the TV news-film release "one of the most effective, yet one of the least known weapons in the public relations arsenal."

Television stations, expanding local newscasts to help fill the evening prime time turned back to them by Federal Communications Commission rulings, have generally welcomed the commercially produced news releases. "The advent of the so-called film handout has given public relations men a sophisticated tool that its advocates say is all but indistinguishable from news that a station's own staff has gathered," the *Wall Street Journal* explained.

A Los Angeles-based news-film production company charges $345 for producing a sixty-second release, plus a $15 fee for each television station to which the news clip is sent. According to the *Handbook of Public Relations*, the cost of covering 100 stations in this manner in 1970 ranged from $26 to $33 per station.

The president of WorldWide Films, headquartered in Metuchen, New Jersey, says that on the basis of a distribution to 300 television stations, the average news-film clip will be used on at least 100 news shows with a viewing audience of around 20 million. Figuring total costs to the client of under $5,000, that works out to about forty cents per thousand viewers—one-tenth the cost of commercial time, or four thousandths of a cent per viewer. Even a postage stamp, he notes, costs eight cents. For optimum acceptance by television news directors, WorldWide recommends a sixty-second news-film clip, in color with sound track.

PR WIRES AND NEWS FEATURE SERVICES

The newspaper equivalent of the television news-film clip service has been around for many years. Derus Media Services' "Editorial Pace" and Associated Release Service, both of Chicago, are two of the more successful feature services. For a fee the news feature organization will distribute a story to weekly and daily newspaper editors all over the country. A one-column wide by seven-inch long matted publicity release sent to 1,000 papers by Derus Media Services, for example, costs $250 plus postage. Associated Release Service's charge for essentially the same service is $260. For an extra fee most feature services will also research and write the article.

The private wire services offer the advantage of more rapid distribution of a story, but to a smaller number of outlets. The PR Wire Service of Boston and New York offers private line teletype coverage of news media in sixty major cities in twenty states from coast to coast for $70 per story serviced. Subscribers to PR Wire pay an annual membership fee of twenty-five dollars. Washington, D.C.'s Chittenden Press Service offers its clients a unique dual service on press releases. For around $30 per release Chittenden will disseminate 400 copies of a press release to the ubiquitous Washington press corps—the largest concentration of news correspondents in the world. Special direct delivery to the offices of all senators and representatives is also available, should anyone be interested.

Of more direct interest to the design professional is the companion service, wherein Chittenden's staff collects and delivers to its clients on a daily basis press releases from forty major government agencies plus the White House. The fee for this service also runs about $30 a month. Inquiries about these and other services may be directed to Chittenden Press Service, 1841 Kirby Road, McLean, Virginia 22101.

MISCELLANEOUS PR TOOLS

In this necessarily skeletal review of a few of the major tools and techniques available to and utilized by the public relations profession, much has been omitted or touched on all too briefly. The self-generated and often self-serving seminar is a tool used by more and more design professionals. These sessions, as they apply to architects, engineers, and planners, will be discussed in the next chapter. Preparation of a press kit by professionals is a subject worth at least a chapter unto itself. Company literature, including internal and employee publications, is another whole field in its own right, as are the more esoteric offshoots such as public relations abroad and political public relations. Planning and evaluating effective public relations programs, along with suggestions on how to select a compatible public relations consultant for your firm must await another book or the reader's own research.

MOUNTAINS, MOLEHILLS, AND MANURE

While publicity, according to Carl Byoir, cannot whitewash a manure pile, nor, according to other authorities, create a mountain out of a molehill, it can cause one mountain or one event to seize on the public's attention, while similar or even more important mountains and events sink into obscurity.

Who has not heard of Pikes Peak, for example? The gold rush of 1858–1859 and its slogan, "Pikes Peak or Bust," helped to get General Zebulon Montgomery Pike's mountain known all over the world, but few are aware of the fact that Colorado has thirty other peaks higher than Pikes Peak.

James Rumsey built the first steamboat and put it into operation on the Potomac River in 1789—at least eighteen years before Fulton's trip up the Hudson River in the Clermont. Fulton is credited with the first "profitable" steamboat—an odd distinction for inventions in the field of transportation. On that basis, Wilbur and Orville Wright would be just another bicycle shop family partnership. It should be mentioned that Fulton's trials were held near New York City and its large resident press corps.

Because Longfellow opted to write about Paul Revere's midnight ride, Israel Bissell is an obscure historical figure in spite of the fact that he rode out on the same night, covered a much greater distance, and played a much more historically significant role than did Revere. Bissell rode for four days to take the news from Watertown, Massachusetts, to Philadelphia, where the Continental Congress was assembled. Revere rode a few hours, was captured and jailed before he had completed his mission. Longfellow may perhaps be excused his historical oversight on the basis that "The Four Day Trip of Israel Bissell" does not have quite the ring of "The Midnight Ride of Paul Revere."

Even the worst history student knows about Garcia and the message carried to him. Who remembers Andrew S. Rowan, the messenger?

A final example of the power of publicity and press coverage— whether planned or accidental—concerns Mrs. O'Leary of DeKoven Street in Chicago and her cow. According to legend, the cow kicked over a lantern and set a small outbuilding on fire, which grew into the great Chicago Fire of Sunday, October 8, 1871. Property loss in the Windy City was a staggering $187 million; 300 lives were lost as a result of the fire.

Few are the persons in the civilized world who do not know about the Chicago fire and Mrs. O'Leary's clumsy bovine. Equally few are the persons who do know about another, much more destructive fire on the same night a few miles from Chicago, the worst disaster of its type in the United States. Only two other U.S. calamities, the Galveston, Texas, hurricane in 1900 and the Johnstown, Pennsylvania, flood of 1889, claimed more lives than were lost in the great fire of Peshtigo, Wisconsin, on October 8, 1971.

At least four times as many lives were lost in the Peshtigo conflagration as in Chicago; more than 1,250,000 acres of pine forests were destroyed.

The Peshtigo fire began as numerous small fires in the tinder-dry forests of northeastern Wisconsin. Early in the afternoon of October 8 a heavy wind came up from the southwest. The flames quickly joined together into a firestorm, working its way northeast into the town. The first realization of danger for most of the town's 2,000 residents came as their homes burst into flames. Many tried to reach the nearby river.

"Some were burned to death within a few feet of the river, some in their houses, some in the woods and some on the roads attempting to escape," wrote one of the few survivors. "Within half an hour, and some say within 10 minutes of the time the first building caught fire, the entire village was in flames. The great sheets of fire curled and rolled over the ground like breakers on a reef."

The local priest described the sight as "nothing but flames. . . . Houses, trees and the air itself were on fire. Above my head, as far as the eye could reach into space . . . I saw nothing but immense flames covering the firmament."

The next day the long overdue rains came and extinguished the flames, but only after they had burned unchecked nearly twenty miles into upper Michigan.

Chicago had many newspapers and hundreds of reporters on the scene. Its fire received millions of words of publicity, while Peshtigo, Wisconsin, and its far worse fire was largely ignored—and is now all but forgotten.

REFERENCES

[1] Philip Lesly, "Communicating with a Segmented Society," *Public Relations Journal*, June 1968, p. 11.

[2] Herbert M. Baus, *Public Relations at Work*, Harper & Brothers, New York, 1948, pp. 135–136.

[3] Donald T. Van Dusen, "Is Publicity a Neglected Art?" *Public Relations Journal*, December 1969, p. 17.

[4] Ibid., p. 17.

[5] Vincent Drayne, "How to Lose an Audience," *Public Relations Journal*, September 1967, p. 20.

[6] Fred Isley, *Stylebook for Press Releases*, rev. ed., Xerox Corporation, New York, September 1970, pp. i–ii.

[7] Mel Snyder, "How to Get the Best in Pictures," *Public Relations Journal*, March 1969, pp. 18–19.

CHAPTER 15

Public Relations: Advanced

A number of public relations-oriented tools and techniques and outlets for various publicity materials have already been covered or touched upon in Chapters 5 and 6: Promotional Tools and Strategy, I and II. These included:

Brochures—the planning, production, and distribution of.
Reprints from domestic and foreign publications and the piggy-backing of same.
Books authored by design professionals.
Film production.
Letters—to register interest in a prospect, to make a proposal, and to thank the potential client.
Job histories—the importance of, for all types of research and reference.
Special events—open houses, speeches, seminars, parties.
Competitions—participation in.

All the above should be considered in their original context as well as in connection with the subjects covered in this chapter.

PLANNING PAYS OFF

Implicit in any discussion of public relations for the design professional is the existence of a thoroughly planned program, based on the firm's generally accepted goals and projected five, ten, even twenty years into the future.   When one is in a profession where a single project may require five to eight years to go from concept to completion, short-range planning for any aspect of the overall operation is shortsighted, probably next to useless, and certainly offers potential dangers and unnecessary conflicts.

A planned approach to public relations should always include the three considerations covered in Chapter 14: evaluation, identification, and communication.   The process may be thought of as designing a kind of road map.   It is a map of the diverse roads to one's public relations goals, including intermediate stops, and like any map is designed to save traveling down a lot of wrong or dead-end streets and becoming trapped in blind alleys.

One public relations practitioner summed it all up in rather appropriate terms:

It is far easier to tackle your public relations problems on a day-to-day basis instead of approaching them in terms of a pre-determined policy. The difference between a planned public relations program and one which is improvised from day-to-day is like the contrast between a magnificent architect-designed home and a bungalow to which an impulsive owner has added first one room, then another, until the formless building sprawls across his land like a freight train.[1]

The results of an internal, rather general, survey of one design firm's public relations program may be of interest and serve to illustrate some of the points to be made in this chapter:

Public relations is a two-sided function, involving external and internal (employee) relations.   The staff is only one of many publics which must be reached, influenced, and convinced by a planned and productive public relations program.   Our clients represent yet another important public.

In implementing and carrying out our public relations program, the following points are offered as guidance:

1. Project the image of our firm as an organization (as opposed to an individual) of broad base and qualifications.   In connection with this point, it is a truism of public relations that to capture successfully the public's imagination a publicity program must be built around *people*. People are news—ideas generally are not.   When people are "sold" or "promoted" or "publicized," so are their ideas.   Example I: History is

written in terms of men.   The American Revolution and the Civil War were fought for ideas—yet history is written about Washington and Lincoln.   Example II: The development of architecture down through the ages is based on ideas—ideas of design, construction techniques, esthetics; but the history of architecture is written about Wren and Wright and Michelangelo.

2. Develop opportunities for exposure in various media, e.g., newspapers, magazines, television, and radio.   Assist key personnel of the firm in preparing speeches, public statements, and articles.   Publicize professional and civic activities of such personnel when appropriate.

3. Set up opportunities for partners, associates, and other key personnel to make public statements, serve on panels, and participate in seminars; then publicize these activities.   The other side of this particular coin is that we cannot be incommunicado when newsmen, on their own initiative, contact us for information or a statement on a subject about which we should be knowledgeable.   In this connection, it is important that all staff members keep the public relations department posted about any approaches made to them by media representatives. It is a legitimate technique of news gathering to whipsaw a source—and we could find ourselves in the embarrassing position of being on both or even several sides of the same question.   This is not censorship, but coordination, and is highly important in any professional endeavor.

4. Prepare articles for magazines on timely issues regarding the profession, public issues of local and national interest, and our own work.

5. Prepare information and obtain publication of our buildings:
    a. As projects (with sketches) in the architectural press and in newspapers.
    b. As completed buildings in all media.

6. Write news releases about new work assignments as clients agree to the release of this information, and for formal ceremonies such as groundbreakings and dedications.   The latter events should be anticipated by the project architect involved, with the partner-in-charge having the final responsibility for alerting the director of public relations. Public relations will then offer all possible assistance to clients in publicizing the project, including the preparation of press kits.

7. Maintain a system for updating the VIP mailing list and institute periodic VIP mailings of information about our office and its activities, utilizing reprints, award announcements, copies of press releases, and any other pertinent material.

8. Project the firm's image to potential employees through regular contacts with universities and professional journals, coordinated with the personnel director.

9. Retain a national public relations agency to assist in projecting our firm nationally and internationally.

10. Promote and execute retrospective exhibits of the work of the firm on local and national levels.

11. Maintain liaison with local, state, and national professional organizations.

12. In the area of internal (employee) relations include the following activities:

 a. Backup services to project teams through photography, brochure preparation, and other pertinent graphics.

 b. Operation and maintenance of a professional library, being always responsive to staff needs.

 c. Maintenance of a photographic library of our work, including still photographs, movies, and slides, particularly covering design development stages.

 d. Regular publication of an internal newsletter.

 e. Social activities, including a Christmas party and other major "employees only" parties, along with minor events. Program and coordinate arrangements for these activities.

 f. Art exhibits. These could include an annual exhibit of staff-produced art, plus art exhibits from local museums and galleries on a loan basis.

 g. Maintenance of up-to-date biographies and photographs of key personnel.

The survey concluded with an excerpt from an article by Paul L. Field in the January 1970 issue of *Business Management:*

> Press releases, as one evidence of the fact that American companies are placing more and more stress on their public communication, are becoming more sophisticated today. It is no longer assumed that a presidential ribbon-cutting for a new plant ranks in interest with a moon landing. Or that the chairman of the board merits headline treatment because he adopted a band of Cub Scouts.

PRESS CONTACTS

"Why Is the Press Sometimes Wary?", an editorial by Neil Gallagher in the September 1965 *AIA Journal,* contained several morsels for thought:

> . . . the newsman needs not renderings but rapport.
>
> He doesn't want the architect to write his story; he needs help in appraising the subject.
>
> He doesn't want to be patronized; he needs a solid basis of understanding.
>
> He doesn't want to be impressed; he needs a set of impressions from the knowing.
>
> He doesn't want to be exploited or deceived; he needs help and guidance in helping the public to perceive.
>
> He doesn't want to be cultivated; he needs communications.[2]

George McCue, the award-winning architectural critic of the *St. Louis Post-Dispatch*, adds: "Some [architects] are aloof, some are poor communicators, some are so busy that they are hard to catch at the moment a newspaperman needs contact."

ETHICAL BOUNDARIES

Some design professionals appear to have almost as strong a fetish against public relations as they do about selling or marketing their services. William B. Chapman, former executive director of the Philadelphia AIA chapter, once covered the practicalities of the matter in these words:

> Like death and taxes, public relations for the architect is a matter of *no choice*. You practice your profession in the public eye, making decisions about buildings and streets and places that affect the homes and industry and education and recreation and commerce of all men . . . decisions that change both the way we live and the quality individually and collectively. You relate to the public in countless ways—and thus, you are going to have public relations whether you like it or not.
>
> The only question left to discuss is whether you're going to have good public relations or bad public relations.
>
> Because, to be part of the action, there is no other choice.

EARLY EFFORTS

Lest anyone think that the preoccupation with public relations for the design professional is of recent vintage, in January 1940 the *Weekly Bulletin* of the Michigan Society of Architects published a lengthy article, "Progress Toward Public Relations." The article carried a condensation of remarks by D. Knickerbacker Boyd to the Philadelphia AIA chapter and the T Square Club at a joint session of the two groups on November 15, 1939. One of Boyd's recommendations at that long-ago meeting concerned the issuance of a series of postage stamps, "including good examples of architecture, engineering and construction—with portraits or the names of architects, and in some cases of the engineers and contractors."

Another Boyd suggestion was somewhat more practical, with a slightly contemporary note:

> Due to the possibility of there being too few building trade workers in the near future, it is suggested that adult classes or lectures be conducted on every phase of the building industry, including real estate, architecture, engineering, contracting, manufacturing, erection and installation. Such courses could be conducted by the combined associa-

tions in the industry and consist either of short or concentrated periods. To these would be invited all local elements in the industry, including workers, especially those who, due to lack of employment, are now out of the industry and occupying positions as chauffeurs, barbers, bartenders, filling station agents, etc.

IMAGE PROBLEMS

If the image problem of the professional designer seems to be a difficult one to solve, consider the plight of the American Wood Preservers Institute. Its members annually spend a sizable sum on public relations in an uphill effort to convince the public that utility poles are a desirable addition to the landscape. The concept of burying utility lines in urban areas has somehow to be made to look bad.

Following loud and organized public opposition to the addition of more utility line towers to the New Jersey landscape in 1972, the utility companies got into the utility pole act. The Public Service Electric and Gas Company of Elizabeth, New Jersey, enclosed an "informational" folder in a mailing to customers in July 1972 (see Figure 15-1).

Part of the problem is that no one really knows what it would cost to put utility lines underground. Obviously, there would be some additional cost. Note that PSE&G quotes $250 for each customer in the third paragraph and two paragraphs later the figure has risen to $550 per customer. It appears that in the first instance PSE&G is talking about *all* overhead lines; in the latter, the reference is only to transmission lines. The sowing of planned confusion is another public relations technique.

Some of the overhead wire opponents wonder how it is possible to locate problems almost immediately in a spaceship circling the moon 239,000 miles away, while it is so difficult for a utility company to find and repair line troubles a few miles from the generating plant. Perhaps earth technology will overtake space technology before visual pollution overtakes us all.

POLICING ADVERTISING

A few years ago a ruling by the Board of Ethical Review of the National Society of Professional Engineers, prohibiting all advertising other than professional cards, was handed down at about

Should power lines go underground or overhead?

Here are the facts.

With today's emphasis on the environment, there is an increasing demand that all electric lines be installed underground and not overhead on towers or poles. However, this would be very costly.

To place all existing lines underground throughout PSE&G's service area would cost approximately $3½ billion. That is far more than the total we have invested in all our generating stations, towers, poles, lines, trucks, equipment, offices, and land combined!

(page 1)

If, **beginning in 1973**, we were to install all planned overhead lines underground, the cost for this one year would be about $400 million more than the cost of installing them overhead. This cost would increase each year and, based on present growth rates and escalating costs of labor and materials, would reach an estimated yearly cost of $800 million by 1980. The $400 million additional cost for just 1973 represents an investment of $250 for each of PSE&G's 1,600,000 electric customers.

Such a phenomenal increase in our installation costs would result in substantially higher costs to customers. This would be a hardship on many families and a deterrent to prospective new industries which could bring new jobs and dollars to New Jersey. This may well explain why there is no requirement in any state in the nation for installing all new or existing electric lines underground.

For example, the added cost of installing needed new transmission lines in Union and Middlesex Counties underground, as requested by some residents, would be approximately $70 million more than if installed overhead basically on existing

(page 2)

PSE&G and railroad rights-of-way. This cost, if borne by approximately 127,000 customers in the 12 affected communities and not shared by customers in other areas of the state, would be about $550 per customer.

PSE&G is currently installing new residential lines underground in subdivisions of three or more homes in accordance with a new regulation of the New Jersey Board of Public Utility Commissioners. The additional cost of burying these lines is borne by the builders.

It costs 2 times as much, or more, to install these residential lines underground instead of on wood poles. But in most cases it ranges as high as 20 times as much to install high-voltage lines underground instead of on steel towers because of the need to use very expensive specially-insulated cables. Other additional costs are encountered in excavation and pavement replacement in urban areas, and the longer time needed to locate, dig up, and repair trouble in underground lines.

A balance must be found between these high costs and esthetic values. For instance, PSE&G has been installing lines under-

(page 3)

ground for years, wherever congestion of overhead facilities would result in operating difficulties, such as in downtown streets. Wherever possible, new high-voltage lines are installed in existing PSE&G or railroad rights-of-way. In other areas, towers are installed with minimal tree clearance, and high points on the landscape are avoided.

We present these facts to you so that you can better understand the complexity of our transmission problem. Please be assured that if we could put all lines underground today without having to raise the cost of electricity drastically, we would do so. But it can't be done! However, we are constantly seeking better ways to deliver energy more efficiently, economically, and without disturbing our environment, and we will continue to do so. After all—the Energy People are Environment People, too!

The Energy People

Public Service
Electric and Gas
Company

(page 4)

Fig. 15.1

the same time the Consulting Engineers Council liberalized the ground rules for advertising in its Guidelines for Professional Conduct.

The NSPE's Board of Ethical Review, in a special report to the board of directors in July 1969, acknowledged the difficulty in determining whether advertising was "non-self-laudatory, dignified and circumspect." The report went on to point out that to follow this course would mean "trying to define, ad infinitum, vague and generalized language on an advertisement-by-advertisement basis." The BER recommended a change in the language of the advertising code and suggested that "the Board of Directors direct the Ethical Practices Committee to prepare a revision of the Code of Ethics to prohibit the advertising of engineering services." The revision was adopted seven months later and letters of notification went out to firms who had been advertising.

FIVE STORIES FOR EVERY PROJECT

With practically every project the design professional has at least five story possibilities. In chronological order they are:

Signing of the design contract.
Bid opening and construction contract award.
Groundbreaking (with or without rendering).
Topping out.
Dedication.

In addition to the plus coverage that truly imaginative groundbreaking, topping out, and dedication activities can bring, there are a number of side-bar (feature-type) articles related to the basic list of stories. The side bars include:

An historic or unusual site.
Construction site activities and events.
Progress stories.
New or unusual design features or construction techniques.

While every job does not lend itself to the "bigger-than, more-than, longer-than, deeper-than" approach, the comparative feature is applicable to many buildings. An example of this type of story

was in the *New York Times* for June 6, 1969, "Trade Center Doing
Everything Big."

> In addition to coils of wire that could be unravelled and stretched from
> New York to Seattle and more steel than is in the Golden Gate Bridge
> and enough air conditioning equipment to cool 15,000 homes, the World
> Trade Center now rising in lower Manhattan will need doorknobs.
>   Forty thousand doorknobs.
>   . . . the six buildings will have 230 acres of rentable space, nine more
> than the 18 buildings of Rockefeller Center, and their demand for elec-
> tricity will be equal to that consumed by a city the size of Schenectady,
> N.Y.[3]

While it helps in this case to be writing about the tallest office
towers in the world (100 feet higher than the Empire State Building), one doesn't have to deal in record breakers to come up with an
interesting comparison story.   This feature about the Hartford,
Connecticut, Civic Center was prepared by the architect's public
relations department and released by one of the investors in the
project.

> Hartford, Conn. [Date]—Imagine a regulation football field completely
> covered with a pile of dirt as high as the _____ Building.
>   If statistics are your bag, that's almost 10 million cubic feet of earth.
>   Hartford, Conn., faces the monumental task of excavating just this
> amount of dirt for its new Civic Center, designed by the Philadelphia
> architectural firm of Vincent G. Kling and Associates.
>   The Civic Center will be an unusual venture of public and private
> funds on the same four-block site in center city Hartford.   The City of
> Hartford is responsible for the construction of an arena, exhibition
> space and parking; Aetna Life and Casualty is financing shops, theaters
> and an office building.
>   James P. Barbour Jr., Director of Project Development for Aetna,
> has computed that the excavation will leave a hole averaging 30 feet
> deep and covering 7½ acres.   Before it's all over, the 350,000 cubic yards
> of dirt removed will add up to 29,200 truckloads.   That translates into
> a 100- by 300-foot football field piled 330 feet deep.
>   Anyone with a hole of the approximate size might contact James
> Barbour at Aetna in Hartford.

The blank in the first paragraph was to allow for customizing
of the story for local newspapers.   The article's writer furnished
a list of forty-seven buildings in as many cities.   All the structures
were approximately 330 feet high, to make the volume of earth
described in the second paragraph come out right.   The first few
entries in the building list looked like this:

City	Building	Height
Akron, Ohio	First National Tower Building	330'
Albany, New York	State Office Building	388'
Atlanta, Georgia	Regency-Hyatt House	330'
Austin, Texas	State Capitol	311'
Baltimore, Maryland	Hearst Tower	330'
Baton Rouge, Louisiana	Louisiana National Bank Building	277'
Birmingham, Alabama	City Federal Building	325'
Boston, Massachusetts	Suffolk County Courthouse	330'
Buffalo, New York	Manufacturers & Traders Trust Co.	315'
Charlotte, North Carolina	North Carolina National Bank Bldg.	289'

To remove all mystery from how such a story is written, an extensive list of the tallest buildings in the United States is found in the *World Almanac*.

Successful exploitation of events at the construction site really is limited only by the public relations counsel's imagination and initiative. Even when the story is written tongue in cheek, as in the next example, it should have a cumulative positive effect.

"Bank Lands Fish Using Bulldozer" was the headline of a two-column feature story, accompanied by a two-column photograph, in the *Dallas Times Herald* several years ago.

This, is an old, old fish story (75 million years) that happened in downtown Dallas.

It is also a documented story and, unlike many fish tales, can be proven.

The story was leaked to the press this week by the Republic National Bank, which caught the fish in its excavation site on St. Paul. (This is unusual because the fish was caught by the bank instead of from it.)

Explaining that their catch was 75 million years old, the Republic National called out the various Dallas news agencies, large and small, and gave them steel fishing hats.

And, at the site, over the loud roar of machinery and the annoying roar of television cameras, a public relations man explained, "This is a cephalopod."

He pointed out that Southern Methodist University said it was a phylum Mollusca of the cephaloda class from the Cretaceous Age.

"We dug it up with a bulldozer," said a workman.

Observing the catch, the press reporters took careful notes and the photographer rushed to get a picture.

Offering helpful information, one photographer explained to another photographer from the other Dallas newspaper, "These things are common. My kids dig them up in the back yard all the time."

He took a picture anyway.

And, a television man tried to take a picture, but the newspaper photographer walked away with the fish.

"And with that a workman walked up with another cephalopod. "We found another one," he said.

It was a good day for cephalopods.[4]

It was not, unfortunately, a good day for the architect, engineers, or consultants, who were not mentioned even once (the cephalopod rated four mentions)—but you can't win them all. This does point up the need for continuing participation in publicity planning and activities by design professionals.

GROUNDBREAKING

Since there usually is just a piece of reasonably open ground to deal with in a groundbreaking, one can think in terms of fairly spectacular—even destructive—events for a ceremony to signal the start of construction. The first sentence was qualified with "usually" because occasionally weather or other acts of God will force a postponement of the groundbreaking until after construction has gotten underway. In one extreme case large tarpaulins covered the first floor structure during the symbolic groundbreaking; the building was about 15 percent completed at the time. Everyone present wore a sheepish expression during the event—all the while trying to pretend that the building wasn't really there.

Sometimes it is possible to make an event out of the beginning of demolition on a site. A story is told about a famous governor of a populous Eastern state who agreed to run the crane for the first swing of the headache ball. The machine was carefully set up ahead of time, naturally, and all the governor had to do was climb into the cab, pull one lever—and the old apartment building being torn down to make room for a large luxury hotel would then have a large symbolic hole knocked off one corner at about the third floor level.

Unfortunately, the crane's regular operator didn't get the boom prepositioned quite right—or the governor pushed when he was supposed to pull. Whatever the reason, the wrecking ball swooped cleanly into the third floor corner of the next building. Since this was a perfectly good office building, the crown was stunned. The governor was very embarrassed and the organizers of the event were petrified. And the owners of the wrongly wrecked building sued everyone in sight.

Accidents can always happen, needless to say, whether it is a site clearing, a groundbreaking, or a topping out, and one is well advised to try to plan for every contingency.

Since that story may have discouraged all but the strong in heart from demolition events, we will return to groundbreakings.

Just about everything that can be done visually with shovels and bulldozers has already been done. Gold-plated shovels, silver-plated shovels, chromed shovels, and beribboned shovels have pretty well had their day as far as the press is concerned. Multiple shovels with a single handle are still good for a picture now and then, especially if a couple of miniskirted models are among the shovelers. It is hard to find a dependable source for the multi-bladed shovels, however—and they can get very costly.

A VIP up on a bulldozer is almost always good for a few laughs —and sometimes a photograph—but a lot of VIPs object to getting their suits and shoes dirty climbing on and off the machines.

Hard hats seem to capture the fancy of most nonconstruction executives. If the hats have been decorated with the wearer's name, the date, and the name of the project, it is even odds that he will take the hard hat back to his office and put it on display with other prized mementos. This fascination with the construction industry on the part of outsiders helps to explain why a conservative businessman will travel miles to an out-of-the-way, unshaded site on a hot summer day to watch another conservative business-man—whom he may or may not know—turn over a shovelful of pre-dug dirt. It also partially accounts for the popularity of the ancient occupation of sidewalk superintending while a building is under construction. More on sidewalk supervisors a little later.

Controlled explosions, watched from a safe distance, seem to be among the most popular forms of groundbreaking. This caption was under a photograph showing balloons floating up toward an airplane, while a cloud of smoke drifted out of the picture to the right:

Marriott found a way to get "big" press coverage of a usually dull event —a groundbreaking. Officials and the press cruised over St. Louis in a jet while conducting groundbreaking ceremonies. When the plane was directly over where a new Marriott Motel is to be built near Lambert Airport, J. Willard Marriott, board chairman, pushed buttons and guests watching through the windows saw the hotel property erupt into clouds of colored smoke and helium-filled balloons. The new Marriott will have 433 rooms when it opens in '72.

Union Oil used much the same idea for a new office building in Anchorage, except that guests were kept on the ground (but well back from the blast area). Hundreds of rubber balls, rather than balloons, were thrown high in the air by the detonation. Presumably, the guests spent the remainder of the afternoon chasing down the bouncing balls.

When a creative PR type is turned loose to devise something new in groundbreakings, the results sometimes rival the best (or worst) of Rube Goldberg's wonderful inventions. Figure 15-2 illustrates the complex, inventive approach, as applied to a groundbreaking in Florence, Kentucky, several years ago.

The accompanying caption explains:

> Starting up a bulldozer with a blowtorch was the novel way to launch construction of a $3.7 million insulation board plant in Florence, Kentucky for the Great Lakes Carbon Corporation of Kentucky.
>
> Mr. D. L. Marlett (left), vice president of Great Lakes Carbon Corporation, and Lt. Governor Wilson W. Wyatt are shown performing a blowtorch "wedding" of the symbol for New Industry with molded lead symbols of two of Kentucky's traditional economic mainstays: Horse Racing and Tobacco. Demonstrating the incombustibility of the Permalite insulation board to be produced in the new plant, the blowtorch melted the lead which ran down the undamaged board to an electrical contact, starting the first site-clearing bulldozer.

The caption continues on for two more rather long paragraphs, but the important facts are in the excerpt. His participation in this complicated groundbreaking stunt was apparently no particular boost for Lt. Governor Wyatt's political career. Running for the U.S. Senate not long after the ceremony took place, he was defeated by Thurston Morton.

It may have struck some readers by this time that no architectural or engineering credits appeared in any of the articles quoted. These stories were selected, in part, for that reason, once again to make the point that the designer must involve himself in all such publicity gimmicks, or suffer the consequences of anonymity.

TOPPING OUT

The following Associated Press release, published in early 1962, has always been one of my favorite top-out stories.

> New York (AP)—The top steel went on a new Park Avenue bank building today and along with it went a fir tree and a basket containing eggs, vegetables, handkerchiefs, corn sheaves, ribbons, a pair of handcuffs and a jar of chicken blood.

Fig. 15.2

The chicken blood is an old Chinese custom—a substitute for human sacrifice—and the handcuffs were the nearest thing the Bankers Trust company could get to the ancient shackles tossed into the Hellespont by Xerxes the Great to punish evil water spirits.

The food is for the steeds of the ancient god Woden, to persuade him to send lightning elsewhere, and the fir tree at the top is an old Norse custom.

The handkerchiefs and ribbons, symbolizing sacrifice, are a more recent method used by construction crews to ward off evil spirits.

A bank vice president, Herman G. Maser, did the research to make sure the new building between 48th and 49th Streets would get an auspicious start.

"We are playing it safe," he said, "and taking no chances."

Quite often, as habitues of top-out ceremonies know, the only decoration is an American flag affixed to the steel. This practice reportedly goes back to the beginning of this century. Where a tall building is involved it is important to make a vantage point available to the press on a nearby building, if possible, where they can see the beam dropped into place. Once in a while a photographer wants to be in position on the building being topped out, for some picture angle he has in mind. In such cases it is best that the attorneys for the client, designer, and contractor do not know about it until they read the evening paper. A release is always advisable anytime a newsman or any other visitor to a project is to ride a construction elevator or walk across temporary flooring twenty to forty stories up.

Sometimes a project lends itself to multiple top outs. Each of the three major theaters in the John F. Kennedy Center for the Performing Arts in Washington, D.C., was topped out appropriately as the last girder went into place on the individual sections. Completion of the steelwork for the Concert Hall, for example, was marked by raising a 10-foot-long steel replica of a bass viol to the top of the section housing the hall.

DEDICATIONS

The dedication is usually the most important and, therefore, the most newsworthy step in a building's progress. The groundbreaking signaled the beginning of construction and the top-out ceremony indicated that things were well along, but the dedication tells the world that the years of planning and construction have come to a successful conclusion and the building is ready for the client to occupy it.

Dedication ceremonies range from the simple to the complicated, from the informal to the very dignified, and from the ridiculous to the sublime. One should be creative in planning a dedication — but not outlandish or undignified just for the sake of being different. If the building is a stable, then a horse show might be appropriate, but a parade of antique cars or a beauty pageant would not normally be a desirable adjunct to the dedication of a college science building.

One approach to the dedication of a library was described in the *AIA Journal* for January 1972:

> The three-day dedication of Alvar Aalto's library (for Mount Angel Abbey) just out of Portland, Oregon, could be called a swinging affair, literally and figuratively, for it all began with a Friday night concert by Duke Ellington who presented the music of Ann Henry in a premier performance of *Pockets — It's Amazing When Love Goes on Parade*. Miss Henry, by the way, who composes music in a popular idiom and has written a Mass and a number of hymns, has done some of her creative work in residence at the abbey.
>
> On Saturday the architects of Oregon and the Northwest had an opportunity to tour the library, and when Aalto found he could not be on hand for the dedication, a panel was assembled to discuss the features

of the building. . . . Later that day, the Portland Junior Symphony presented a concert in the library, and Professor Richard W. Southern, president of St. John's College, Oxford, England, lectured in the abbey church.

Sunday's festivities got underway with a concert by the Lewis and Clark College Choir and Abbey Schola, followed by the "official" dedication and blessing of the library, with addresses by the head of the Benedictine Order from Rome and the Finnish Ambassador to the United States — and just for good measure a presentation of a collection of books on Judaism.[5]

TIMING

There is no need to rush a dedication. The inexperienced client may assume that the contractor's completion date is sacred and inviolate — but there is no excuse for an architect or engineer accepting a completion date at face value. Strikes, weather, and foulups in material deliveries are a few of the familiar causes of delays which can push deadlines past the point of believability. Even after a client has taken beneficial occupancy of his building there remains the final cleanup and finishing touches to the landscaping. Be sure the building is ready before setting up dates and sending out invitations.

While no one will guarantee good weather, it is a good idea to check out the long-range forecast for dates under consideration — especially if it is to be an outdoor event.

Martin Brower, in an article on dedications in the *AIA Journal* for January 1972, pointed out:

A building dedication can be one of the most useful means of bringing a completed project, as well as client and architect, to the public's attention. For the client, nothing is as illustrative of the organization's progress, stability, strength and concern for its employees and for its community as a well-conceived new building. The same is true for the architect. A well-executed project glowingly illustrates his firm's abilities to meet the needs of the client and community.

It only makes good sense, then, for the architect to assist the client when it comes to dedicating a building rather than shying away from the responsibility. By providing this additional service, he also gets a chance to further his own public image.[6]

Dedications can be simple affairs, involving no more than a few words of welcome from the client, followed by brief introductions of the architect and contractor. Or they can be elaborate, lengthy, and costly affairs.   It usually depends on the client's wishes, objectives, and the size of his promotion budget.

Dedications often feature a symbolic opening of the structure, with a ribbon cutting the most-used symbol.   Cutting a ribbon is about as trite as shovels for ground-breakings, so if something related to the client's business can be adapted as a symbol, the better the chances are of interesting news photographers in covering the event.   At the dedication of a Wish-Bone salad dressing plant in Independence, Missouri, several years ago, a 36-inch, hand-carved wooden wishbone on a wide green and gold ribbon stretched across the front entrance was broken to signify the formal opening of the new processing center.   Since the ceremony took place in former President Harry Truman's hometown, Mr. Truman was the guest of honor at the dedication.

Which brings us to the question of who should be invited to groundbreakings, top outs, and dedications.   When the event is important enough and sufficient space is available, some clients issue a mass invitation to the general public.   This can be particularly effective when a series of guided tours of the new facility can be arranged.

Some obvious invitees are the client's representatives and the architect, engineer, and general contractor.   Consultants and major suppliers and subcontractors may also be included.   Government officials—local, state, and as far up the line as one can reach—are usually invited, along with whatever celebrities can be assembled.   Media representatives are not only invited but catered to.

Martin Brower, in "How to Dedicate a Building," says:

> A good way to interest the press in the new structure is to give a press preview before the dedication.   This can be done by inviting the press to tour the building with the architect, either as a group or individually. It is up to the architect and his client to make the tour interesting by pointing out how the building differs from other buildings or by indicating its special functions.   Then, when an article is submitted at the time of dedication, the press will be more receptive.
>
>   The best way to obtain coverage of a project, regardless of its importance, is to have press materials prepared by a professional public relations person and submit this to the news media.   It is important for the architect to stay close to the preparation of client-prepared materials if he wishes to be included and to have the project well described.   He can do this by supplying the basic materials or even by handling the entire preparation, with professional help.   A basic press kit for a building dedication should include:

1. An overall press release on the building, describing in news or feature style the program and the solution, the use of the building, its location, its size, its materials and its special features.
2. One or more good 8 x 10 black and white photographs.
3. A brief, basic fact sheet, listing the bare facts of owner, architect, size, materials, etc.
4. A concept piece written and bylined by the architect.
5. A brief background on the client and the architectural firm.[7]

A copy of the client's dedication remarks (hopefully brief) is also sometimes included in the press kit.

Many buildings have more than purely local interest, so a mailing should be prepared for the out-of-town press, construction and architectural publications, and for all trade and consumer journals having even the remotest connection with the client's business. The client's PR department should have a current list of publications covering his own field, but the conscientious public relations counsel will run the client's list against such sources as *Bacon's Publicity Checker, Standard Rate and Data Service,* and the *Newspaper and Magazine Directory* published by Luce Press Clippings, Inc. Luce's directory covers almost 4,000 trade and consumer magazines—including such nonhousehold names as *Peanut Journal & Nut World, Macaroni Journal,* and *Earnshaw's Infants and Childrens Review*—along with the 1,900 daily and 8,200 weekly newspapers read and clipped by Luce for its customers.

The dedication in October 1971 of the reconstructed London Bridge in Havasu Lake, Arizona—described by the *New York Times* as "an extravagant splash of ancient pageantry and modern press-agentry"—gives some indication of what is possible with an extremely generous budget and a client with a flair for promotion and gimmickry.

A highlight of the weekend-long series of parties, tours, and varied entertainment, laid on to put the guests in the proper celebratory mood for the actual dedication, was a lavish banquet for 800 guests. The dinner, reported to cost more than $500,000, featured lobster and roast beef, the same menu used for the original dedication of London Bridge in 1831 by King William IV. This would be a hard act to follow by any standards, but the next morning the Lord Mayor of London and the Governor of Arizona pulled on a red ribbon to release thousands of colored balloons and hundreds of white pigeons. At the same instant appeared skydivers, rockets, and a large hot-air balloon towing a huge dove of papier-mâché. Adding to the strange mixture of old- and new-world customs and cultures, the University of Arizona marching band took to the streets of Havasu Lake, led by leaping, well-tanned, baton-twirling majorettes. A visiting British newsman was quoted as saying: "It's all quite mad—it's a supergimmick and could only happen in America. Only an American would think of investing that much in something as crazy as this."

CONSTRUCTION SITE ACTIVITIES

On-site public relations mileage can be obtained from a number of other activities of lesser importance than groundbreakings, top outs, and dedications. The safety fence around the site can be decorated in all manner of ways. Provisions for the traditional peepholes at several levels for the use of sidewalk superintendents should never be overlooked. Official Sidewalk Superintendents Clubs, with membership cards and special newsletters, have been organized at some job sites. The job sign should be placed for maximum visibility and contain all appropriate credits. A three-dimensional model of the building may be part of the job sign, to show the public what the completed building will look like. Tenants' names are sometimes placed on a special section of the fence or on the construction bridges over the sidewalks.

COMMUNICATIONS

There are several important publics to keep in touch with as the building plans progress and construction gets underway. These include other property owners near the site and the client's own employees and stockholders. Mutual Of New York (MONY) sends out this letter to families, business firms, and their employees who live and work in close proximity to the firm's building sites:

> Probably you have read in the newspapers that MONY is about to start construction of its new office building. While "sidewalk superintendents" seem never to be fazed by the noise and dust, we know that these things may be sources of annoyance to you.
>
> We would like to ask your indulgence during these days of our construction work. Our architects and contractor will be making every effort to carry out the project with the least possible inconvenience and annoyance to you.
>
> We feel sure that when our new building is completed, it will be a

BUILDING BRIEFS

Fidelity Mutual Life   Girard Bank

GOING UP?

Have you always dreamed about having an elevator ready
to serve you when you need it? Not ten minutes later!!
The elevators in the Fidelity Mutual Life Building may
not be exclusively yours, but they will provide more
flexible service than any other elevators in the City.

HOW COME?

There will be four banks of Westinghouse Mark IV, high speed elevators
continaing a total of 24 elevators. Each elevator will hold 21 people.
Of the 24 elevators, two will be special service elevators to do the
heavy work. They will carry up to 6,500 pounds at 500 feet per minute.
Each of the four banks serve approximately one-quarter of the building
which results in fewer stops for each elevator. The elevators serving
the upper half of the building will travel at 1,000 feet per minute
while those elevators servicing the lower half of the building will travel
700 feet per minute. That's how come you will have the best elevator
service in town.

MOVING STAIRS

Since there are three levels below ground - basement, concourse and upper
concourse - there will be moving stairs to transport people quickly from
the concourse level to the main building lobby. Two flights of moving
stairs will be provided to make the trip from the concourse level to the
lobby as fast and as comfortable as possible. Hence, no matter which
subway or which railroad you use to come downtown, you can always come
into the building from the concourse level. An added feature of the
moving stairs is that they both can move in the same direction at once
depending upon traffic.

The moving stairs will be available for your use during normal business
hours, however, stationary stairs will be available for your use after
hours. These stairs take you from the concourse out to the street on
the north side of Girard Plaza.

All in all, you will find that there are more elevators per rentable foot
in the Fidelity Mutual Life Building than any other building. No one in
Philadelphia has any faster elevators. Aren't you glad you're with the
best!

No. 5   November, 1970

Fig. 15.3

worthy contribution to the progress of your community. After that, we
will be quiet neighbors. In the meantime, your neighborly understanding will be very much appreciated.[8]

The letter is signed by MONY's president.

Special internal newsletters are one way of keeping the client's own employees, executives, and stockholders abreast of developments in a new headquarters building. Figure 15-3 is a reproduction of one of the series of "Building Briefs" newsletters published for the employees of the Girard Bank and the Fidelity Mutual Life Company in Philadelphia. Figure 15-4 is the cover sheet of the first of several employee bulletins written by the General Motors public relations department during the construction of the General Motors Building at 767 Fifth Avenue in New York City. The introductory paragraph of the General Motors newsletter explains:

> This is the first of a series of bulletins for General Motors employees in New York City intended to answer your questions about the new General Motors Building and inform you about public transportation to the new area, the location of departments and activities, general services (dining, elevators, communications, etc.), neighborhood facilities, the schedule for moving and other important subjects.

SIDEBARS

A PR counsel for a design professional is not restricted to stories about his client's buildings. A little applied ingenuity can turn up features such as the following, issued as a release early in 1957:

BABEL ON 67TH STREET — AN ARCHITECTURAL WONDER

Note to 1968 architectural graduates: If you consistently earned top grades in design, are looking for an interesting office in which to work, and have been to Mongolia or speak Albanian, you might send an application along to the New York office of Edward Durell Stone. The internationally famous architect would like to complete his own staff set of languages and countries.

While world travel and fluency in foreign languages is not a normal requirement for registration as an architect, it might appear to be a condition of employment in the far-flung operations of architect Stone.

What began recently as a routine survey of staff language abilities in the Stone organization turned up a real linguistic bank, even for New York City. The internal poll was taken to locate potential escort-translators for the many visitors who find their way to Stone's imposing town house office-residence at 7 East 67th Street, as well as to find additional in-house translators for the relatively large volume of foreign language mail received there.

The 146 employees reporting to date (out of almost 200 in three offices on both coasts) show complete speaking and reading fluency in eighteen modern languages, plus usable abilities in six additional tongues. Practically all European languages are represented (exceptions: Finnish, Greek, and Albanian), along with more esoteric languages such as

January 2, 1968   Bulletin #1

GENERAL MOTORS CORPORATION

<u>ANNOUNCING</u>

<u>"A NEW VISTA ON FIFTH AVENUE"</u> --

<u>THE GENERAL MOTORS BUILDING</u>

*   *   *   *

(This is the first of a series of bulletins for General
Motors employes in New York City intended to answer your
questions about the new General Motors Building and inform
you about public transportation to the new area, the location
of departments and activities, general services (dining,
elevators, communications, etc.), neighborhood facilities,
the schedule for moving and other important subjects.)

*   *   *   *

Sheathed in white Georgia marble and rising 50 stories into
the Manhattan skyline, the new General Motors Building at
767 Fifth Avenue will be a place about which you will be
able to say proudly: "That's where I work".

The exterior marble soon will be completed, making the
building's dignified design and serene simplicity more
evident to all who are watching it take shape as truly,
"A New Vista on Fifth Avenue".

The "vista" will not only be on Fifth Avenue. Viewed from
any side of the 84,000-square-foot block -- Madison Avenue,
58th or 59th Street -- the GM Building will be an urbane
complement to its famed setting.

But enough of the really unnecessary sales talk! We all
know that famed architects Edward Durell Stone and Emery
Roth & Sons have designed a fine building -- but just how
big will it be?

The building will be the 12th tallest in New York City with
the 50 stories extending 705 feet. For comparison, the
Empire State Building's 102 stories and TV tower go up
1,472 feet. The closest comparison with its fellow tall
sentinels on the New York City skyline are the 707-foot,
52-story Union Carbide Building on Park Avenue at 47th Street
and the 50-story Metropolitan Life Insurance Building which
rises 700 feet from Madison Avenue between 23rd and 25th
Streets.

The first two floors of the General Motors Building will
contain a total of 84,000 square feet. The 48 tower
floors will each have about 31,600 square feet -- enough
room to park about 270 Chevrolets.

Fig. 15.4

321

Arabic, Korean, Malayan, and Persian. Counted only once in the total is a Chinese draftsman who can handle five of his native dialects.

While the great majority of the architectural staff—71 percent—was born in the United States, twenty-five other countries of birth are represented in the employee roster. Some, like Ed Stevens, son of veteran Moscow newsman Edmund Stevens, were born in foreign countries but have U.S. citizenship through their parents.

Another question in the survey asked for a list of countries lived in or visited. More than ninety countries and all fifty U.S. states were listed. The UN currently has 123 members, so few major powers have been missed by the ubiquitous Stone employees. Somehow, in all their travels the far-ranging design staff has missed Mongolia, Bhutan, Chad, and Pitcairn Island—and a few other spots.

That this particular employee group should be so well traveled and spoken is not quite so unusual as it might at first appear. Besides working with such internationally oriented American clients as General Motors and PepsiCo, Inc., Ed Stone has designed structures all over the world in his thirty-six-year architectural career. In addition to heavy domestic commitments, some fifteen overseas projects by his firm are now underway in Saudi Arabia, Panama, Lebanon, Pakistan, and other distant lands.

The list of colleges and universities attended by the Stone staff likewise reads like an international Who's Who in Education, from schools such as the University of Bratislava (Czechoslovakia), Warsaw Politechnika, the Sorbonne, and the University of Capetown.

A Department of State official, who directs some of the foreign visitors to Stone's office through State's Reception Center in New York City, commented in a recent letter to Stone: "I doubt that many universities turn up fluency in 24 languages. It is an interesting commentary on the rich and varied background of your associates."

Meanwhile, if you plan to get an architectural degree in 1969, it's not too late to begin learning Thai or Urdu.

SEMINARS

An increasingly important tool for architects, engineers, and planners is the self-generated seminar, on practically any subject. It is perhaps a sign of the growing aggressiveness of some design professionals that they no longer are content to wait for invitations to sit on panels sponsored by a professional, commercial, or management organization; they now sponsor and organize their own conferences, seminars, and panels.

In Chapter 3 we saw how a New Jersey-based design firm was able to capitalize on its seminar on community colleges. With that seminar an unqualified success, members of the firm were established as specialists in the community college field and the commissions began to come in.   Once some actual experience was gained and a few of the college projects were completed, the commissions increased to the point where an occasional one had to be refused.

A few years later, the principals of a Midwestern A–E firm predicted that vocational-technical or career-education schools would be a worthwhile specialty to pursue.   This firm had done some of the pioneer work in voc-tech school design, but had not been active in the field in recent years.

A series of seminars was determined to be the best vehicle for promoting the firm's interests.   A traveling team of experts from the A-E firm, plus a representative of the Office of Education, HEW, in Washington, was assembled and materials were produced.   The seminars were held in almost a dozen cities—from Anchorage to Houston to Boston—and attracted thousands of educators.   So many, in fact, that the firm soon realized that it would have a problem in following up on all of the potential jobs represented.   That is a nice kind of problem.

MISCELLANEOUS

Design firms located in major television and motion picture production centers such as New York and Los Angeles can occasionally cash in on the entertainment industry's insatiable appetite for stage properties (props).   Some corporations go after motion picture screen and TV tube exposures of their product or service so avidly that they must be convinced the effort pays off.   TWA, for example, set up its Film and Television Promotions department in 1951 to act as a distributor of authentic TWA movie props and a variety of full-scale aircraft mockups.   The department also owns an extensive library of aircraft-oriented film clips and worldwide establishment shots.   If a client needs a view of a 707 landing at the airport in Hong Kong at dawn, TWA will pull the appropriate clip from its library.   Since 1951, TWA's insignia and equipment have been highly visible in more than 300 motion pictures, 1,000 television shows, and hundreds of educational and documentary specials, plus many non-TWA sponsored commercials.

No one is suggesting this type of ambitious approach for a design firm, but a few letters to the person in charge of props, describing what can be provided on short notice, could bring some satisfying national exposure for renderings, photographs, and models not otherwise occupied with anything more productive than gathering dust.

The public affairs director of a New York-based architectural firm sent the following letter to all New York television stations a few years ago. Since the title of the department in charge of props varies from station to station, it is a good idea to verify the name of the person in charge, his title, and the proper name of his department by telephone in advance of writing:

Mr. Art Edwards, Director
Scenic Design Department
WNBS-TV
3 East 67th Street
New York, New York

Dear Mr. Edwards:
A recent request from one of the television networks prompted me to write to you.

An architectural rendering of an apartment building was required as a prop for a live show and the stage designer called to ask if we could help him. Obviously, we could and did.

Should you ever have need of a rendering, photograph, or model for one of WNBS-TV's shows, please feel free to check with us. Chances are that we will be able to come up with something to meet your need.

I realize that such a prop is not an everyday requirement, but from my own experience I know that when you do need an unusual prop, it's usually an emergency request.

If the situation is such that credit can be given our office for the model, rendering or photograph—and such credit is desirable from our standpoint—rental charges would be only the legal minimum of $1 per day. Otherwise, the rental charge would run very little more, since we are not in the prop rental business. We would expect WNBS-TV to pick up the prop and return it to our office.

Sincerely,
Gregg Jordan, Director
Public Affairs
Petlewany, Gallunzi & Bloy, Architects

Brief "FYI" memos to distant newspaper editors and news directors of radio and television stations, well in advance of an event, are acceptable substitutes for personal visits to promote coverage. Telephone follow-ups a few days before the event is to take place are always advisable. The subject of the FYI memo may be almost anything newsworthy. As an example, this memo, sent to the city desks of the two newspapers in St. Louis, resulted in many additional inches of coverage for the architect involved:

October 3, 1967
TO City Desk
    St. Louis Globe-Democrat

FYI, Edward Durell Stone, design consultant for the Busch Memorial Stadium, will be attending the World Series game on Sunday, October 8, as the guest of James P. Hickok, Chairman of the Board, First National Bank in St. Louis. This will be the first occasion for Mr. Stone to see his design concept in use (we're holding thumbs that he doesn't get caught in a traffic jam on one of the ramps).

Mr. Stone will be sitting in First National's box. I thought his presence might provide a photo for game color or a small sidebar—which is the reason for this memo.

John Doe,
Executive Assistant—Public Affairs

The memos resulted in a two-paragraph "Special Visitor" mention in a pregame story in the *Globe-Democrat* on October 6; a four-paragraph sidebar feature in both the *Globe* and the *Post-Dispatch* in the October 9 papers as part of the game coverage; a two-column photograph in the *Globe* on October 10; and a special fourteen-paragraph feature, "Stadium Designer Lauds City's Vision" in the *Globe* for October 11.

CLIPPING BUREAUS

At some point in a firm's development one of the principals usually decides that the services of a clipping bureau are necessary to follow and evaluate external publicity efforts. There are several services in the United States; some are national in their publication coverage, others are regional or local in scope. One of the largest clipping services suggests that new clients cover the following points with their clipping bureau:

The Where: Details on the location of the company, or hometown of the individual. Include location of subsidiaries, sales offices, etc.

The What or Who: Information on what the firm or individual does. What makes it newsworthy? What is being published?

The Keys: A list of trade names and generic terms to be used in news or feature stories. This is to give the clipping service's readers every possible way to watch for the order—and extra chances to find clippings, however a paper or magazine may choose to print the story.

The Luce Press Clipping Service suggests that the following information can be learned from studying press clippings:

1. Editorial acceptance accorded your publicity and product news releases . . . regionally, nationally, and even internationally.

2. What new products and services are being developed and marketed in your field.

3. Public and editorial attitudes toward your firm or organization, people, policies, and services.

4. Personnel changes and appointments in your customer, prospect, or competitive organizations.

5. Factual case histories of your products or services in use.

6. Proposed and impending regulations and legislation having a bearing on your profession or business.

7. What your competitors are offering and claiming, where and at what prices.

8. Pros and cons on controversial subjects having a direct or indirect bearing on your company's activities and services.

9. Complete information on labor markets, labor conditions, contract negotiations, and settlements in your own and related industries.

10. Lists of births, deaths, marriages, and other specific data; date and location of fires, robberies, disasters.

CONCLUSION

The best public relations, one veteran has noted, is foresighted, not defensive—positive, not negative. It relies on truth, not subterfuge. It is deeds accomplished and communicated, not persuasive words alone.

And, finally, consider the question once posed by PR pioneer Edward Bernays: "What is the most difficult part of our work?" His answer: "The client."

REFERENCES

[1] Lee H. Bristol, Jr., *Developing the Corporate Image*, Charles Scribner's Sons, New York, 1960, p. 294.

[2] Neil E. Gallagher, "Why Is the Press Sometimes Wary?" *AIA Journal*, September 1965, pp. 41–42.

[3] © 1969 by the *New York Times* Company.   Reprinted by permission. June 6, 1969, p. 45.

[4] "Bank Lands Fish Using Bulldozer," *Dallas Times Herald*, May 9, 1962, p. 36.

[5] "Asides," *AIA Journal*, January 1972, p. 4.

[6] Martin A. Brower, "How to Dedicate a Building," *AIA Journal*, January 1972, p. 27.

[7] Ibid., p. 29.

[8] John P. Brion, "How to Build Goodwill from a 'Hole in the Ground,'" *Public Relations Journal*, March 1967, p. 20.

CHAPTER 16
A Mood of Change

A BRIEF REVIEW

In Chapter 1 we saw that the design professions are beset not only with shortages of trained personnel, but also with confusion in the minds of clients and among the general public and other professions as to exactly what is the contemporary function of architects, engineers, and consultants. This external bewilderment is accompanied by increasing and mutual encroachments on traditional design roles by fellow professionals and consultants. Excerpts from Royal Barry Wills's 1941 book, *This Business of Architecture*, demonstrated an early awareness on his part of the techniques of selling professional design services. Not until the first Canons of Ethics were adopted by the American Institute of Architects in 1909 did the design profession find it necessary to establish rules governing professional practice and ethics. The subject of professionalism—in general and specific terms—was also covered in detail.

Chapter 2 was basically a review of generally accepted marketing principles and psychology, with heavy emphasis on effective and persuasive communications. Some of the tested guidelines for productive selling were enumerated. Many of the principles discussed in Chapter 2 were illustrated and amplified on in the chapters following.

By Chapter 3 we were ready to get down to specifics on a business development program for the design professional. An introspective, objective look at the professional's own firm was the recommended first step, followed by the selection of one person to head up the sales effort. Suggested promotion budgets and establishment of prospect contact files were outlined, accompanied by illustrations and examples. The importance of making thoughtful analyses of the current practice mix, determining desirable directions of a firm's potential growth, and establishing the reasons for lost jobs were all dealt with at length, as was the setting up of branch offices.

Chapter 4 covered the identification and qualification of prospects, an extremely complex and important subject. Internal and external sources of continuing, productive information about prospective clients were listed, along with basic materials and books for a Business Development Reference Shelf.

In Chapters 5 and 6 the more important promotion tools for selling design services were listed and discussed, together with their application and the elements of a general promotional strategy. The tools included brochures, reprints, books, motion pictures, television, correspondence, job histories, lectures, and seminars. There is considerably more to producing a brochure than one of the principals writing an introductory statement to a motley collection of job photographs and floor plans. Tips and suggestions for dealing with brochure consultants and printers were given, along with representative client reactions to brochures. Authorship of articles and books by design professionals was strongly recommended. Large sections of Chapter 6 were given over to examples of letters to express interest in a project, and proposals and other correspondence related to selling a design firm to a prospective client. *Architectural Record's* Specification Outline was cited as a guide for setting up and maintaining job histories. Two examples of how to explain design services to clients were given.

What to do before, during, and following the presentation or interview was the general subject of Chapters 7 and 8. Completion of the qualification process and stepped-up intelligence gathering, including site visits, were discussed in Chapter 7. Contingency planning, the operation of Murphy's law, and "murder sessions" took the balance of that chapter.   VIP treatment of potential clients and general guidelines for making presentations were the primary points covered in Chapter 8.   "Show biz" has its place in every presentation.

Practical politics was the subject of Chapter 9, in which lobbying, Washington representation, campaign contributions, and the organization and support of political parties were explored in depth. Some of the scandals involving the design profession and political contributions were discussed, along with the difficulties of policing such contributions.   How, when, and in what amounts to support candidates accounted for the rest of Chapter 9.

Chapter 10 dealt with governmental clients on local, state, and national levels.   Preparation and distribution of the U.S. Government Architect-Engineer Questionnaire—Standard Form 251 was explained, and a few specialized government markets were listed. The chapter closed with a brief discussion of payoffs and kickbacks to government officials.

The rationale behind associations and joint ventures was explored in Chapter 11, together with the how, when, and why of associating.   Responding to Requests for Proposal's (RFPs) was also covered with examples.

In Chapter 12 we considered the pros and cons of pursuing overseas work—an interesting if often frustrating field of endeavor. A number of sources of job leads and assistance were given, together with some dos and don'ts of foreign operations.   Overseas work, it was emphasized, is not for every firm.

Taking a piece of the action was the subject of Chapter 13—with profit possibilities and potential pitfalls sharing about equal coverage.   Reasons for the increase in designer-entrepreneurs were examined and the importance of surrounding oneself with competent advisors was stressed.   Readers were advised to learn how to analyze real estate developments.

Since it is practically impossible to divorce public relations from business development activities, two chapters—14 and 15—were given over to an exposition of basic and advanced public relations. Chapter 15 emphasized planning for effective PR and gave numerous examples of promotion possibilities and story ideas based on the design and execution of projects.

This brings us to the present and final chapter.   As a general rule, readers seem to expect the last chapter in a book of this type to be a combination of review and summation, exhortations and prognostications—or at least that is what publishers tell authors the readers expect.   A kind of look-into-the-future-where-do-we-go-from-here?

Hopefully, enough elements of these requirements will be included in this wrap-up chapter to satisfy both readers and publisher. Accompanying and expanding on the author's forecasts will be the thoughts of a few contemporary practitioners and clients on the future of design and construction, plus several ideas of the author's which did not seem to fit logically into any of the preceding chapters.

THE BUSINESSMAN'S WRAPPINGS

Archibald Rogers, FAIA, one of the more articulate American designers, once described a design professional as "an artist surrounded by a professional wrapped in a businessman."   Architect Rogers added, "We are therefore a sort of missing link—a triphibian that must be understood and survive in three worlds."   Part of the third of Mr. Roger's worlds—the businessman's wrappings— is mostly what this book has tried to be about.

As William Dudley Hunt, Jr., pointed out in his book *Total Design:* "The slightest attention to the history of architectural practice, even in its earliest eras, will reveal architects selling their services to pharaohs, kings, wealthy merchants and others.   Yet the myth that architects need not sell their services persists even down to the present.[1]

The AIA prefers "project development" as a euphemism for sales. In the chapter on public relations in the *Architect's Handbook of Professional Practice* is found: "Project development is the actual searching out of prospects, responding to inquiries regarding commissions, making the professional presentation, and following up until the owner has made his selection."[2]

As we all know, advertising by individual design professionals is generally prohibited by all of their associations.   The AIA member

- MAY NOT buy an advertisement in his name in any print or broadcast media;
- MAY NOT join with other Architects in buying an advertisement un-

less the ad is AIA Chapter-approved and includes the names of *all* Chapter members or those in the immediate geographic region, regardless of whether or not they contribute to the cost of the ad.[3]

Well and good—but what is the poor professional designer or consultant to think when his direct competition—faced with no such prohibitions against advertising—places advertisements such as this one for Walter Kidde Constructors, Inc.?

> . . . a Boise-Cascade Company—designs, engineers and constructs manufacturing plants, offices, schools, hospitals and transportation facilities throughout the world. The company is a leader in developing and implementing such innovations as construction management. Offices are maintained in major U.S. cities and abroad.

Little comfort in this and similar cases will be derived from the AIA's statement that "advertising by individual professionals *does not* make good business sense."[4]   The explanation? "Since (unlike retail sales or manufacturing) the unit cost of architectural service is not reduced by any quantity production that might be caused by an increased demand for services, advertising costs necessarily would be passed on to the client in the form of higher compensation or reduced service."[5]   Think about that for a while.

CHANGING TIMES

But the times, they are a changing—and some veterans in the design profession expect most, if not all, of the strictures on advertising and self-promotion to be relaxed or voided completely in the next five to ten years.

Since change is one of the few constants remaining to us in today's complex world, we tend to search for signs of it in many places.   Over the 1972 Labor Day weekend the *New York Times* heralded two diverse straws in the winds of change blowing on the construction industry.

On Friday, September 1 the *Times* described the construction of a five-story Inter-Continental Hotel in Mecca, Saudi Arabia. Since only Moslems may enter Mecca, the construction supervision, by a twenty-man team of non-Moslem French and West German technicians, was accomplished from a control tower eight miles from the construction site.   The Europeans monitored the 700 Moslem workers on a fifteen-camera closed-circuit television hookup and relayed instructions to the Moslem foreman by walkie-talkie.

Other details of interest about this unusual project are the marble balconies outside each guest room, with markings in the floor to show guests where to stand to face the Kaaba (sacred black stone) in the Beit Allah (house of God) when they pray. Because Moslems are forbidden hard liquor, the hotel bar serves only nonalcoholic beverages.

The other *Times'* Labor Day weekend item appeared on Sunday, September 3—and was not calculated to perk up a hard-working, underpaid designer. It was the story of thirty-nine-year-old Tom Dowd, an operating engineer foreman at the World Trade Center construction site in downtown Manhattan. In 1971 Dowd was reportedly paid at least $94,000 for supervising his fellow operating engineers—whether they needed supervision or not and whether he was actually on the job or not. Some $76,000 of Dowd's total income was for overtime. The *Times* calculated Dowd's total take from his six-year stint on the World Trade Center project at more than $500,000. Dowd, who did not care to be interviewed, agreed only that he had "made a good buck last year."

Elsewhere in the story was the mention that the costs of the Trade Center complex have risen as steadily and inexorably as the twin 110-story office towers. The original cost estimate of $350 million had grown to more than $700 million, with many months of construction remaining.

Steeply escalating construction costs are no longer real news, of course. The exceptions are generally when the overruns become truly astronomical, as in the New York State office complex in Albany or the new FBI headquarters in Washington, D.C. In mid-1972, at 20 percent completion, estimated costs of the latter project had soared to $107.7 million from the original estimates of $47.7 million.

ONE CLIENT'S REACTION

Clients, as discussed in Chapter 13, have begun to notice and comment on the more obvious faults and shortcomings of the design and construction industries. The publisher of *Engineering News-Record* took the unusual step of commemorating McGraw-Hill's move to its new headquarters in Rockefeller Center with the following editorial:

After 40 years in the big, green building that for many years has been McGraw-Hill's trademark, *Engineering News-Record* has moved to a new New York City home. The new McGraw-Hill building, at 1221 Avenue of the Americas, is part of Rockefeller Center.

Like so many other buyers of building construction, McGraw-Hill has suffered many of the outrages which, in recent years, have become part of the building package. Because of rivalry, mindless vandalism or calculated malice we have suffered:

- Smashing of corner beads in freshly plastered walls
- Spillage of 10,000 gallons of hydraulic fluid for two truck elevators
after the filler cap was smashed
- Damage to the central water chilling system
- Obscenities scrawled on freshly painted surfaces
- Cutting of wires in prewired cafeteria equipment
- Cutting of 40 telephone cables
- Theft of construction materials

The price of all this will be about $2 million.

There must be a more efficient way of getting a building built.

Eugene E. Weyeneth, Publisher[6]

GROWTH INDICATORS

Increased efficiency is called for in many other areas besides building construction, of course. Consider that more than 80 percent of all solid waste in this country is still disposed of in open-pit dumps —a disposal system developed by the armies of Julius Caesar. About the only major change in waste handling over the last several centuries is that we now carry it to the curb in cans and sacks instead of just heaving it out of a window into the street.

That there have been significant improvements in some areas of our life style cannot be denied. In just five decades life expectancy has increased some 50 percent, while the work day was being reduced by one-third.

In the first quarter of 1972 the United States began producing goods and services at an annual rate of more than $1 trillion, the first nation to pass this economic milestone. There are predictions that we will achieve a $2-trillion gross national product by 1980.

In the late fifties the United States became the first major power with a service economy, when less than half of those in the nonfarm labor force were classified as factory or manual laborers. By 1970 those in service industries outnumbered goods producing employees by more than two to one.

Clearly there will be more people to serve [in tomorrow's economy] and they will be better educated and more affluent than the market of the '60s and certainly more discriminating. It is doubtful, for example, that they will settle for the ticky-tacky houses and neo-penal apartments that were produced in such great numbers in the 1960s.

By 1980, it is expected that about 80 percent of all housing work will be done in the factory.[7]

George Romney, then Secretary of the Department of Housing and Urban Development, was not quite as optimistic as *Business Management;* his prediction in the early days of Operation Breakthrough was that "at least two-thirds of all housing production in the United States will be factory produced by the end of this decade."

Both forecasts now appear to have been less than justified. By mid-1972 many modular housing builders were trampling over each other to get into bankruptcy courts. When Sterling Homex, a leader in the infant factory-produced housing industry, entered its petition for a Chapter 10 Bankruptcy Act proceeding, a ten-month loss of $26.3 million was reported.

The *Wall Street Journal*, in an article on modular home units headlined "Haunted Houses," pointed out that Sterling Homex was plagued by problems of "production, transportation and marketing, or, in short, every aspect of the business."[8]

The *Journal* story also offered these comments on the fast-sinking factory housing industry:

> Modular, or factory-built housing, to be sure, remains an intriguing idea—producing modules or boxes on a factory assembly line, then transporting them to a building site and swiftly putting them together into a house or an apartment building. Thanks to the factory, builders can avoid the delays of weather and the costs of local labor on the eventual homesite. Thanks to the assembly line, wage costs should be relatively low. Thanks to both, quality control should be easy. It sounds like an efficiency expert's dream.
>
> Yet, as is now starkly clear to a number of bloodied builders, anguished analysts and ill investors, it isn't at all what it sounds like. A large factory means high fixed costs and requires high volume. Moving modules can cost $1 or more a mile, and laws prohibiting nighttime transportation of modules can limit the effective marketing area sharply. Erection can be delayed interminably by local zoning red tape and a welter of conflicting building codes clutched in the hands of skeptical building inspectors.[9]

One quotation in the article would seem to furnish an important key to a basic problem of assembly-line housing construction. A mobile home builder stated: "Little things can kill you on a modular production line. Somebody always wants a different kind of door or upgraded windows or different carpeting." In other words, people—the ultimate judges—will continue to fight against regimentation and for individuality, even while all of the efforts of the building industry and their government are being directed toward achieving lower-priced housing for the masses.

THE PEOPLE MARKET

Since people and their desires will be one of the identifiable constants in the years ahead, let us take a look at this factor.   The U.S. Census Bureau estimates that by the year 2020—fewer than 50 years from now—our population will be somewhere between 307.4 million and 447 million.   Now a range of some 140 million persons is admittedly a considerable one, but many new factors and variables have crept into the business of population forecasting.

A few months after the Census Bureau delivered itself of the above population estimates, the Federal Power Commission added a sobering footnote with a prediction that the nation's electricity needs will quadruple by 1990, possibly requiring a cutback in nonessential consumer usage to help ease the effects of power shortages.

We have always felt it is regrettable, since federal officials insist on making these long-range surveys and predictions, that they do not all use the same cutoff year.   The Census Bureau talks about 2020, while the FPC will only go as far as 1990.   Other bureaus use the year 2000.   As if in answer to this unspoken complaint, the Census Bureau early in 1972 released a list of states expected to show a population gain of more than 35 percent by 1990—thus matching up power shortages with population excesses.   In alphabetical order the state growth leaders are:

Alaska	Connecticut	Nevada
Arizona	Delaware	New Hampshire
California	Hawaii	New Mexico
Colorado	Maryland	Utah

This list could be helpful in making decisions on expansion, opening of branch offices, and concentration of marketing activities.   At the other end of this demographic scale are three states in danger of registering a net loss in population by 1990: North Dakota, South Dakota, and West Virginia.

TECHNOLOGICAL CHANGES

That time may occasionally be compressed where technology is concerned was one of the subjects covered in the commencement address given by Ben H. Bagdikian, former assistant managing editor of the *Washington Post,* at Northwestern University's Medill School of Journalism, in June of 1972:

> Developments that once took decades [now] happen in years or even months. This was certainly true of the technological developments that helped shape our environment. It took about 200 years to go from the invention of gunpowder to its practical use (if "practical" is really the right word). It took 112 years for photography, 56 years for the telephone, 35 years for radio, 15 years for radar, 12 years for television, 6 years for the atomic bomb and 4 years for the transistor.

If we can agree that technological changes will have an increasingly important effect on environment, design, construction, and demand over the next twenty-five to fifty years, what are progressive design firms doing to prepare for these uncharted waters — and what are the dimensions of some of the changes we might reasonably expect?

WILD-BLUE-YONDER THINKING

Some firms — and their number is growing — have established long-range planning groups. Most of these think-tank committees are made up entirely of an office's senior executives; others draw on a panel of outside experts from related fields. The planning committee of Woodward, Clyde & Associates, the San Francisco-based soils engineering consultants, has traditionally been charged with the firm's "wild-blue-yonder" thinking and planning. The results of the group's periodic skull sessions are passed along to the board, which reviews all recommendations and decides which new markets to pursue and how much to invest in each.

Arnold Olitt, Woodward-Clyde's retired vice president in charge of new business, has been chairing business development seminars for other design professionals for many years. According to Olitt, perhaps the most-asked question in these marketing sessions is, "Where are the new markets — and how do you determine what they are?"

WC&A, for example, believes the burgeoning environmental

field will be an important one for the firm's particular talents. The possibility of constructing more facilities underground would require expertise in tunneling and soils analysis and mechanics. Related to subsoil developments is submarine expansion (and both are related to pessimistic predictions about surface land shortages before the end of the century). WC&A's "wild-blue-yonder committee" suggests that ocean-floor living may become a reality, which will require such consulting work as submarine soil mechanics for structures and studying the effects of such buildings on currents and marine ecology. Recovery of test bores from the ocean bed is a subject of current interest at WC&A. One logical avenue of investigation is offshore well drilling techniques.

Olitt also points to airport development as providing future jobs for many firms. The FAA forecasts that air travel will triple by 1980 and the present federal airport subsidy program of around $280 million a year calls for a total expenditure over the next few years that works out to more than twice the amount spent on airports between 1946 and 1969. It is no secret that airport design largely follows developments in aircraft by the manufacturers. Lockheed, for instance, announces that it is putting out a heavier airplane; this new craft will certainly require beefed-up runways and possibly new terminal configurations for loading and unloading passengers and freight. WC&A's staff divides up the top airport and aircraft production executives and regularly interviews them to keep abreast of developments.

Woodward-Envicon, Inc., a subsidiary consulting firm, in two years grew to five offices staffed by some ninety people. W-EI, founded to take advantage of a growing market, consults on bacteriology and pollution and makes impact studies for new power plants.

Olitt adds that he formerly spent a lot of time calling on the power industry because of his conviction that more and more utilities will be switching from fossil fuel to nuclear power.

An interview with Olitt in *Engineering News-Record* a few years ago provided further insight into how a professional new business developer approaches his job.

SELLING SERVICES

Defining his best customer as the repeat client, Olitt described that aspect of marketing in terms of percentages. "If an engineering firm is good, it will keep 80 percent of its clients. The 20 percent it loses is what the guy in my position has to replace. In addition, if a firm's determined to grow at an annual rate of 10 percent, that's 30 percent of the volume of business I have to shepherd through the front door.[10]

Eschewing prepared, memorized spiels for marketing contacts, Olitt stressed his firm's ability as a problem solver based on broad experience.

The National Society of Professional Engineers now has several taped television programs for rent on the subjects of selling professional services and public relations for the professional. Olitt is featured on a 2½-hour tape, "Marketing of Engineering Services," available for a rental charge of $30. Sony and other manufacturers make the playback equipment (Model AV 3600 or equivalent for EIAJ ½-inch format) which should be locally available for rental. Produced by the Professional Engineers in Private Practice section of NSPE, the tapes may be rented by anyone.

Some of the larger developers also have blue sky committees, similar to that of Woodward, Clyde & Associates. A large investor such as the Connecticut General Life Insurance Company must not only know where its future development will go, but what kinds of markets it will have to satisfy.

Dave Eggers, of the Eggers Partnership in New York City, explains that his firm uses a variation of the committee approach to business development. They operate with eleven basic committees, each composed of partners, associates, and top level technical staff. Each committee has continuing responsibility for watch-dogging one particular building type such as hospitals, office buildings, schools, or government work. The groups are headed by senior partners most experienced with the particular specialty involved.

Committee meetings are scheduled at least once a month—more often if there is heavier-than-usual activity in the building type. At the meetings both new and old prospects are reviewed. A side benefit, not mentioned by Eggers, of this type of internal organization for sales is to give the firm's designers a ready source of information about special problems in connection with a building type; e.g., a staff member involved with a hospital design could tap the committee on hospitals for information and advice on solving a problem he has encountered. This could help to overcome the profession's tendency to redesign the wheel every time around.

CHANGING MARKETS

Another example of changing markets and clients is found in public school design and construction. State governments, in many cases, have been gradually taking over funding and construction supervision responsibilities from city and county school boards. This has effectively added new members to school selection committees, whether or not the state's education and public works departments are physically present at the interviews. As this trend develops it could mean that designers and consultants will really be chosen in state capitals, rather than by local school boards. The past record of state designer selection on the basis of political considerations and the graft, red tape, and bias seemingly inherent in some state governments is not encouraging in this respect. But as more and more local funding is funneled down from state and national governments—always in answer to the pleas of local officials, we should point out—the important decisions increasingly will be made at the sources of the funds. Local boards and authorities will be forced to accede to the dictates of those "paying" for the projects.

One developing market was discussed in Chapter 13—the leisure time industry, particularly as related to the development of franchises. Hotel and motel clients—even owners of camping grounds —should be important sources of new business in the coming years. The same will hold true for all types of stadiums and convention centers, along with facilities for continuing education, whether on or off campus.

A greater emphasis on extending health care to all citizens should insure a continuation of growth in every area of the medical care field: hospitals, clinics, nursing homes; and medical, dental, and nursing schools. Certainly related to the importance of this client type is the relatively recent emergence of tax-exempt bonds to finance such community projects as nonprofit hospitals. Any design professional who cannot discuss intelligently the pros and cons of tax-exempts with a potential hospital client is well advised to educate himself in the subject. The public finance department of the First Boston Corporation, 20 Exchange Place, New York City, publishes one of the several excellent booklets on tax-exempts.

Hopefully—perhaps with the spur of new government underwriting and outright subsidies—the seventies and eighties will see a turnaround in the cutbacks in research and development by industry and government. Laboratories, testing facilities, and other related structures will then return to the designer's practice mix.

Anything related to the field of "law and order" is another area to observe, investigate, and become educated about. Detention and correction facilities, juvenile centers, courts buildings, police stations, and training facilities for law enforcement personnel will all draw greater attention from local, state, and federal governments over the next ten to twenty years. Increased financing for planning studies, design, and construction of such projects can be safely predicted.

Forward-looking firms are already applying for associate and sustaining memberships in such organizations as the International Association of Chiefs of Police, as one means of insuring a flow of reliable information about new trends and requirements in this field.

GSA'S NEW TRENDS

The General Services Administration, probably the largest single manager and owner of space in the world, is attempting to strike out into unexplored areas in construction, ownership, and occupancy. In 1972, GSA switched its emphasis to purchase contracts, with expectations of conserving time and funds, increasing local tax revenues, and creating new job opportunities. This was made possible by passage of the Public Buildings Amendment of 1972, which authorized GSA, for a three-year period, to contract with private firms and associations to finance and build projects for federal occupancy. Ownership of the buildings passes to GSA at the end of the contract term (up to thirty years), but during the period of private ownership the structure is subject to local real estate taxes. The government tenants pay rent or an equivalent user fee into a revolving Federal Building Fund. The fund is for maintenance and repair and some new construction. The GSA Administrator, Arthur F. Sampson, made several predictions about the effects of the new legislation:

1. A first-year increase in local tax revenues of more than $22 million.
2. Almost 200,000 new jobs in construction and nonconstruction industries.
3. Better, cheaper, and more efficient buildings for government agencies through savings in construction time, use of new building tech-

niques, and avoidance of the traditional time-consuming, multiple reviews of building projects where GSA is the direct client or represents another agency.

Under consideration by the GSA is a recommendation from the construction industry that future federal buildings be designed for multi-use occupancy, with provision for residential and commercial tenants along with the government agency occupants. This approach will require enabling legislation, which may well be forthcoming.

Commissioner Sampson told the participants in the 1971 Building Team Conference in Detroit, ". . . as an owner I will demand discipline from a very undisciplined construction industry." The industry has been warned by the world's largest single client.

SPELLING OUT THE CHANGES

It seems appropriate to close this chapter and conclude this book with some thoughts found in *END70,* the student publication of Louisiana State University's School of Environmental Design. William R. Brockway, Baton Rouge architect, was writing on "The Future of Architecture" for *END70*'s readers, but practically everything he wrote applies to engineers, planners, and specialized consultants as well.

> Today there can be little remaining doubt, even in the mind of the dullest observer, that the profession of architecture is undergoing serious change. We see the signs all around us; in the development of new classes of clients; in the encroachment of systems analysis and theorists into environmental design fields; in the plethora of editorial copy advocating such things as expanded services, social involvement, advocacy planning, total design and many, many other concepts which were mainly observed in the breach only a few years ago.
>
> The mood of change is in the air. Not yet formalized, the ferment in our profession is part and parcel of the unrest taking place in society as a whole, and therein lies a very real danger. We must learn to differentiate between that which is of real practical value and that which is spurious.
>
> As author William Richard Ewald put it: "For most of the privileged, the old, the middle-class, the middle-aged, change may be coming too fast. For many of the young, the disadvantaged, the black, the poor, the scientist, the professional, change may be coming too slow. But for those whose job is to manage change, to build for it, or to investigate it, the rate of change may be more challenging than it is frustrating." As architects and engineers we fit into the latter category.

Since the advent of the computer, the traditional architectural design process of random choices of alternatives has become obsolete. Most architects are not equipped, either by education or by temperament, to perform their function in a computer technology. Unfortunately, there are others who are, and these are the whole tribe of systems analysts, construction management firms and even industrial giants such as General Electric and Disney, who are now designing buildings, cities and industries that should have been designed by architects. The only trouble is, they can do it better than we can.

The message is clear. If architecture is to survive as a profession, we shall have to do a number of things:

1. Develop meaningful design solutions for the problems of society and our environment.
2. Become conversant with the strict dollars and cents economy of the investment world.
3. Incorporate the mechanical arts of systems development into our design processes.

This is what we face and this is what we must do. The future is bright for young architects coming into the profession, if they will but see and subscribe to these disciplines of the future. It is all too easy to be led down blind alleys, to demand instant gratification of demands for reform. . . . But those who persevere and develop solutions to these problems puzzling the profession today will be the master builders of tomorrow.[11]

REFERENCES

[1] William Dudley Hunt, Jr., *Total Design: Architecture of Welton Becket and Associates*, McGraw-Hill Book Co., New York, 1972, p. 93.

[2] *Architect's Handbook of Professional Practice*, The American Institute of Architects, Washington, D.C., 1971, chap. 8, p. 7.

[3] Ibid., p. 8.

[4] Ibid.

[5] Ibid.

[6] Reprinted from *Engineering News-Record*, June 29, 1972, p. 5. Copyright McGraw-Hill, Inc. All rights reserved.

[7] "The New Life Style of the '70's Reorders Priorities." Excerpted from p. 28 of the October 1971 issue of *Business Management* magazine with permission of the publisher. This article is copyrighted. © 1971 by CCM Professional Magazines, Inc. All rights reserved.

[8] James C. Hyatt, "Haunted Houses," *The Wall Street Journal*, Sept. 7, 1972, p. 1.

[9] Ibid.

[10] "Marketing Methods Pay Off for a Top-Ranking Engineering Firm," *Engineering News-Record*, Aug. 13, 1970, p. 20.

[11] William R. Brockway, "The Future of Architecture," *END70*, May 1971, p. 8.

Index